FROM SOLO TO S
BUILDING A SUSTAINABLE CONTEN

Natalie Marie Dunbar

NEW YORK   2022

"In *From Solo to Scaled*, Natalie Dunbar offers content strategists an informed potpourri of tips, practices, and tools to help them mature and grow impactful content teams."

—Lisa Welchman,
author of *Managing Chaos*

"If content strategy has ever left you feeling lonely, *From Solo to Scaled* will be your new best friend. It's generous, fun, and absolutely packed with practice-building wisdom."

—Sara Wachter-Boettcher,
author, *Content Everywhere, Design for Real Life*,
and *Technically Wrong*

"This will be an instant classic among the CS core library. It builds on those books that have gone before it and leads the practitioner further on their journey. Hard to do, and you managed it and made it look effortless (even though I know it was not)."

—Amy Mihlhauser,
content strategist

"This book is a fantastic guide for anyone who is trying to navigate through the fog of establishing and scaling a content strategy practice. I love how Natalie cuts through the chaos and ambiguity and offers practical, actionable advice and frameworks to follow. As I continue to scale up the practice at my company, this book will definitely be a valuable resource."

—Andy Welfle,
coauthor of *Writing Is Designing*,
Head of Content Design at Adobe

"If you're often the only 'content person' in the room, learn the frameworks you need to articulate content strategy—and your value—across your organization."

—Justin McKinley,
content director, Fortune 100s and startups

"This no-nonsense book with real-world examples is truly a blueprint for building and sustaining a successful content strategy practice. It's aspirational *and* immediately actionable. I can't wait to recommend it to my clients!"

—Meghan Casey,
owner at Do Better Content Consulting
and author of *The Content Strategy Toolkit*

"Wow, this is a comprehensive, relatable yet no-frills manual for success for content designers and strategists of any level. It's really a cheat code."

—Aladrian Goods,
Content Design Manager, Intuit

"I've had to build and scale a content strategy practice from scratch, and I can tell you that it would have been 100% easier if I'd had Dunbar's book in my hands."

—David Dylan Thomas,
founder, CEO at David Dylan Thomas, LLC,
author, *Design for Cognitive Bias*

From Solo to Scaled
Building a Sustainable Content Strategy Practice
By Natalie Marie Dunbar

Rosenfeld Media, LLC

125 Maiden Lane

New York, New York 10038

USA

On the Web: www.rosenfeldmedia.com

Please send errata to: errata@rosenfeldmedia.com

Publisher: Louis Rosenfeld

Managing Editor: Marta Justak

Interior Layout: Danielle Foster

Cover Design: Heads of State

Illustrator: Danielle Foster

Indexer: Marilyn Augst

Proofreader: Sue Boshers

ISBN: 1- 933820-57-8

ISBN 13: 978-1-933820-57-6

LCCN: 2022932841

Printed and bound in the United States of America

For my son, Dakota, my mom, Joyce, my sister, Barbara,
and in memory of my father, Ted

HOW TO USE THIS BOOK

Who Should Read This Book?

This book is for content strategy practice builders, and specifically it's for practices that are being built as a part of, or as a service to, a digital or user experience-focused department or team. Perhaps you're the only content person in the room, or you lead a small content team that you're looking to grow to meet demands, or maybe you're a UX or DesignOps leader tasked with building a content strategy practice from the ground up. Whatever your situation, this book will be your constant companion and guide you during what can sometimes be a lonely process.

What's in This Book?

If you're looking for an actionable plan to help you build your content strategy practice, you'll find it in this book, starting with the Content Strategy Practice Blueprint introduced in Chapter 1. These are the building plans you'll use as you lay the foundation for your practice.

Chapter 1 teaches you how to lay the groundwork to support the first component of the practice blueprint—Making the Business Case. All of the remaining components of the practice blueprint ladder up to support this first crucial element, and are covered in Chapters 2 through 5. These chapters give you details on how to create a sustainable practice that you can later scale to meet the demands of your clients or organization.

Chapter 6 provides you with methodologies for maintaining a strong practice core so that if and when you decide to scale your practice, you'll have the scaffolding in place to build high, but safely.

Chapters 7 and 8 cover the concepts of retooling, i.e., taking an inventory of the tools you've used to build your practice with to see if you need to repurpose or augment them as you prepare to scale. You'll also learn about scaffolding that can help you on your journey to content strategy maturity and the stages of growth within your agency or organization.

Chapter 9 teaches you how to present the work of the practice to leadership at the appropriate levels, using language and concepts that leaders understand. You'll also learn how to address the needs of leaders through the lens of content strategy.

Finally, Chapter 10 is a checklist you can use to inspect your practice-building efforts to ensure that you've covered all of the stages and phases outlined in the blueprint and in the supporting chapters.

This book also includes:

- **Nuts and Bolts:** Tidbits of information, such as definitions or additional resources that will be helpful as you build your practice
- **Power Tools:** Expanded resources and imaterials you can add to your practice-building process
- **Notes from the Job Site:** Personal anecdotes and stories from my experience as a content strategist
- **Notes from the Field:** Interviews and stories from other voices in the field

What Comes with This Book?

This book's companion website (https://rosenfeldmedia.com/books/from-solo-to-scaled-building-a-sustainable-content-strategy-practice/) contains a blog and additional content. The book's diagrams and other illustrations are available under a Creative Commons license (when possible) for you to download and include in your own presentations. You can find these on Flickr at www.flickr.com/photos/rosenfeldmedia/sets/.

FREQUENTLY ASKED QUESTIONS

What do you mean by "From Solo to Scaled?"

This book is meant to help you build or augment content strategy practices of various sizes. You'll benefit from the approaches in this book if you're the lone content person in your business (a solo practitioner), or the manager of a small to medium-sized content team looking to grow or scale your operations to meet an increase in demand. Chapters 1 through 5 teach you the steps for building practices of any size. You'll learn how to maintain the core of your practice in Chapter 6 (also appropriate for practices of varying sizes). Subsequent chapters (Chapter 6-9) focus on sustainability and scaling for when you know you're ready to grow.

Will this book teach me what content strategy is and how to do it?

No. But Chapter 1 includes a list of resources that you can use to learn more about the what and how of content strategy. With that said, this book is written from the point of view of building a content strategy practice that has a UX focus, or one that may be part of a larger DesignOps organization.

What definitions of content strategy are you using to base the practice-building methodologies and approaches that you reference in this book?

There are many definitions of what content strategy is and is not. The sample Content Strategy Community of Practice Charter in Chapter 2 includes these operating definitions from Kristina Halvorson's *Brain Traffic* blog:

- Content strategy guides the creation, delivery, and governance of useful, usable content.
- Content strategy means getting the right content to the right people, in the right place, at the right time.

- Content strategy is an integrated set of user-centered, goal-driven choices about content throughout its lifecycle.[1]

While the blueprint included in this book could apply to building a variety of practice types, the focus here is *specifically* about establishing *user experience-focused* content strategy practices.

What are some key milestones in the practice-building process? Or to put it another way, when will I know I'm succeeding?

How you define key milestones that lead to success depends in large part on what kind of organization you're building your practice in and when it's appropriate for you to engage leadership. You'll learn more about this in Chapter 9.

I'm not a content strategist, but I really think my agency or business needs to invest in building a practice. Can this book help me make the case with my managers or leaders?

Absolutely! Anyone who is a champion of content strategy will find this book useful. Walk through the practice blueprint in Chapter 1 to understand what it takes to break ground. Then decide if you're the right person to do so. If that's not you, you'll need to do a bit of legwork to find the content champions among you. When you do, hand them this book. Give them time to sit with it, and if you can, pledge your support in every stage of the process where appropriate. It may take some time—and for the person or persons tasked with building, the road may be a bit lonely. Follow the blueprint and even if you have to revisit or repeat a few steps, you'll find success.

---

1   Kristina Halvorson, "What Is Content Strategy? Connecting the Dots Between Disciplines," *Brain Traffic* (blog), October 26, 2017, www.braintraffic.com/insights/what-is-content-strategy

CONTENTS

FOREWORD

In 2009, a Google search for "content strategy" returned around 8,000 results. There were two books about it on Amazon. You could send connection requests to all 30 people on LinkedIn with the title of "content strategist." And if you looked *really hard*, you could dig up four, maybe five blogs devoted to the topic.

I know these factoids because that was the year I was trying to do research for my book, *Content Strategy for the Web*. That was also the year I did, in fact, send connection requests to all 30 content strategists on LinkedIn and discovered that I wasn't alone: there were other folks out there who wanted the same thing—get the wider UX field to embrace content strategy as an essential practice. So we all started talking, and the conversation quickly picked up steam. Meetups were held. Books were written. Conferences were born. UX teams got excited, content strategist jobs were posted... and voilá! Everyone everywhere understood the strategic value of content, and all was right in the world. The end.

Ha! Ha! Ha! Not really. Things were still hard. Things *are* still hard. Many organizations out there have leadership that still views content as a commodity—easy to come by, easy to publish—and don't dedicate meaningful resources toward content strategy, content design, or UX writing. (We know these organizations because their websites suck and their apps are hard to use. I am just saying.)

However, good news: there are *also* organizations waking up to the fact that **content can make or break a user's experience**... and they're ready to get serious about content strategy.

Of course, there's a massive gap between "seriously interested" and "seriously invested." That's where you come in. Your job is to help your team (or client, or leadership) transform their thinking and processes so that content strategy becomes part of the very fabric of how business gets done. No big deal, I know. But it is achievable. And *From Solo to Scaled: Building a Sustainable Content Strategy Practice* will help you make it happen.

Natalie's book is by no means a magical, fail-safe formula for scaling up a content strategy practice—no organization is the same, and your path will be yours alone. It's also not a fairytale with a happy ending where all things content play nicely together, no matter who owns them or where they sit. However, it is a gloriously informational and empowering work that our field has needed for actual decades. Natalie presents a "blueprint" for change that doesn't require months of banging your head against the wall trying to get people to *get it*. It's a pragmatic guide to navigate the path of growing a sustainable content strategy practice grounded in shared principles. It is a game-changer, and I'm so grateful it's here.

So. If your personal endgame is better content for everyone, then you're holding your new favorite playbook. Go forth. Get started. Natalie has your back.

—Kristina Halvorson,
founder of Brain Traffic and author of *Content Strategy for the Web*

INTRODUCTION

How do we find more people like you?

I was asked this question by the project manager of a digital experience team at a small, but scrappy, multicultural advertising agency. I'd been brought in as a contract content strategist—the agency's first. The question was asked just after I'd submitted content strategy client deliverables to a hard-to-please healthcare client, and just as agency management was negotiating the scope of work to include another client project—a utility company.

Ever the comedienne, my reply was "Could you be a bit more specific? I cover a lot of demographics."

My cheeky response was not only a feeble attempt at humor, but also was meant to cover up the sheer panic I felt as I considered what was *really* being asked: *Can you help us find another content strategist and build out a team? Oh, and while you're at it, can you also help us establish a content strategy practice within the digital experience team so that we can offer its services to current and potential clients?*

Um, sure!

I hadn't realized it at the time, but after the agency added content strategy to its list of digital experience capabilities, what had started off as a single project with a single client grew into a portfolio full of client opportunities, all to be handled by a solo strategist—me—at least until I could find someone else like me.

Still, as excited as I was, it seemed like it would make sense to stand up an actual practice first—complete with processes, procedures, and deliverables—in order to ensure sustainability *before* we brought another content strategist into the mix.

From Solo...

At that time, our digital experience team consisted of four visual designers, a UX lead, an information architect, and a content strategist (me again), as well as a few developers. There wasn't a procedure for incorporating content strategy into the design and UX process, so inadvertently I stepped on more than a few toes as my colleagues were getting their first exposure to content strategy. Basically, I was trying to do the work without having any structure or scaffolding in place.

Eventually, I found myself in a conference room with my colleagues from visual design, user experience, and information architecture—and later, project management and development—surrounded by whiteboards and equipped with a fresh set of multicolored dry-erase markers and a fresh supply of sticky notes. After a few weeks of work, we'd not only developed an end-to-end process framework that seamlessly included content strategy, but we'd also identified a basic blueprint for building a sustainable content strategy practice within the agency.

A few years later, I moved on to a large healthcare company where I was brought in to help build out and co-lead a content strategy practice with two other colleagues. However, both of those colleagues left the company within weeks of each other for different opportunities. So that left *me* to lead the six content strategists who remained on the team. And I also had to figure out how to begin approaching the painstaking work of integrating our processes with other disciplines in the company's sizeable experience design team, with the ultimate goal of building a content strategy practice that was both scalable and sustainable.

Seems like a dream opportunity, right?

I should mention here that my initial exposure to content strategy was during my tenure as a senior writer for an online directory company. Our director brought in a team of contract content strategists from a large digital agency, ostensibly to help teach us the tools of the trade as we expanded our digital portfolio. We desperately needed their expertise and guidance as we attempted to assess the content we had, both quantitatively and qualitatively, and *then* to figure out what the heck we should do with that content—if anything at all.

Instead, those contractors closed rank among themselves, and our in-house content team never had the opportunity to integrate with them or learn from them. I remember they'd discuss things like "content inventories and audits," and "performing gap analyses," but I don't recall ever receiving any deliverables from those efforts. Ultimately, we were left to figure things out on our own, through lots of trial and error, reading books, and cobbling together resources found online.

My practice of content strategy has definitely evolved over the decade, thanks to the work of many authors whose work informed my development, which was mostly self-guided after that initial exposure. And still, just as there isn't *one* type of content strategist, there also isn't *one* single way to do content strategy, and there certainly isn't one *single* way to build a content strategy practice.

...to Scaled

So when I was offered the opportunity to be the content strategy practice lead at that healthcare company, all I had to go on was my experience building that small practice at the ad agency and the rough practice-building blueprint and process framework I'd sketched out with the help of my agency colleagues.

The big question was: *Could I scale it?*

There were more than a few bumps and bruises along the way—including what felt like a protracted turf war over the title of *content strategist* and a lack of alignment about the process and approach to content strategy with our marketing partners. Still we were able to establish, grow, and sustain a larger user experience-focused content strategy practice based on what I have called the *Content Strategy Practice Blueprint*.

That blueprint is composed of five main components, that together with several guidelines and principles provide a foundation for building a content strategy practice that is suitable for a team of one—a solo practitioner—and can also accommodate building a practice to be scaled within a mid-sized organization or expanded even further to help sustain a content strategy practice built within a larger enterprise.

No matter what the size of the practice you are in, you'll find useful and actionable information in every chapter here for building practices of every size, from solo to scaled.

CHAPTER 1

The Content Strategy Practice Blueprint

I'm fascinated by buildings: single family structures, high-rise dwellings, and especially office towers. As such, I've always had a healthy curiosity about the construction process. For example, Figure 1.1 shows a Habitat for Humanity building that I worked on. From the initial breaking of ground to the completion of a building's façade, I find comfort in both the art and order of construction—how foundations support columns, columns support beams, and beams support floors. When the building plans are followed as written, every element comes together perfectly to create a strong structure that is capable of withstanding natural elements like wind and earthquakes.

FIGURE 1.1
I worked as a volunteer on this building with my UX team.

In my career as a content strategist, I've heard colleagues speak about "standing up a team," or "standing up a practice." There was familiarity in the concept of building a figurative structure that had a specific function or purpose. And, of course, that familiarity stemmed from my fascination with buildings, so the construction metaphor made sense to me.

That metaphor also reminded me of one of my favorite books, *Why Buildings Stand Up*, by Mario Salvadori. Before writing and content strategy became my full-time job, I worked in various roles in residential and commercial real estate. All of those roles exposed me to various phases of building construction and tenant improvements, and reading Salvadori's book helped me understand construction and architecture in an engaging way.

The familiarity I felt when hearing the phrase "stand up a practice" in the digital experience world often stopped short of the *idea* of the building metaphor. For example, practices were "stood up" with no attention to order. Foundations were poured before soil tests were completed, often resulting in skipping the addition of the footings that might be needed to support the foundation, or in the case of the practice, doing the work to ensure that the practice followed the necessary processes to create digital experiences that met the needs of users as well as the goal of the client or business. And *inevitably*, the structure—or the practice—began to crumble.

And sometimes those practices failed completely.

From the Ground Up

Having had the opportunity to build an agency-based content strategy practice from the ground up, and later expanding and maintaining an existing practice within a mid-to-large-sized organization, I began to see that failures often happened because steps crucial to supporting the structure had been skipped. Or perhaps the structure had been compromised because the framework used to build it—if one was used at all—couldn't withstand the constant stress of tension and compression.

When I started to think about what caused these seemingly strong practices to crumble—I returned to the building and construction metaphor to look for possible answers. That's because it's sometimes easier to, er, construct a mental model that's more tangible than the nebulousness nature of digital information spaces.

If the building metaphor still feels a bit weird to you, then try this: think of the last time someone asked what you did for a living. If you're a UX practitioner, or if you collaborate with members of a UX team, you've likely experienced the feeling of the listener's eyes glazing over as you tried to explain the concept of user experience—or as I once saw it described, "making websites and apps stink less." Then think of what might happen if you described the user experience using a more relatable metaphor, such as one of the following:

- The internet is a space.
- A website or mobile app is a destination within that space (and in the case of websites, a space complete with its own address).
- The work you do helps people avoid getting lost in that space.

In keeping with this theme, now imagine that the opportunity that's immediately in front of you—that of building a UX-focused content strategy practice—is a pristine plot of land. Provided you have a solid plan and the right materials and tools, this untilled soil is ready for you to break ground and to stand up a healthy content strategy practice.

So this figurative plot of land you've been given needs someone—you—to till the soil and prepare the space for a structure to be built. And the creation of the plans for that structure, as well as sourcing the building materials and the tools you'll need to build it, has also fallen to you.

Lucky for you, this book is your blueprint.

NUTS AND BOLTS TENSION AND COMPRESSION

In construction, *tension* happens when building materials are pulled or stretched. In the process of *standing up* (or building) a content strategy practice, tension can happen when you are asked to take on tasks that pull you away from the core functions of the practice.

Compression happens when building materials are pushed against or squeezed. As you're building your practice, compression may present itself as pushback from departments outside of your immediate cross-functional team. You'll find more details on how the concepts of tension and compression can impact your practice in Chapter 3, "Building Materials."

Department, Team, Practice: What's in a Name?

So, why the focus on *building a practice*? Why not focus on creating a new (or expanding an existing) content strategy department, or focus on hiring a team of content strategists? First, the focus on buildings and structures is intentional. That's because I've learned that for the work of content strategy to succeed as the function that happens *within* the structure you are building, it must begin with a sense of permanency—a firm foundation. Ask any content strategist how many times they've been asked "when the content strategy was going to be done," and how many times they had to explain in response that "the content strategy is *never* done"—that content has a lifecycle, from content creation to archival; that there will most assuredly be legacy content that will need to be maintained in some shape or form; and that the creation of new content (or the addition of newly curated content) starts the cycle all over again.

Content departments and content strategy teams often sit in a variety of places within an organization or agency, including marketing or some variation of digital or user experience. There are also content strategy teams embedded in different organizational functions, such as customer care; or teams that support a specific product or feature, such as video content; or those that are aligned with a single line of business within an organization, such as in a healthcare organization where practices support individual and family products, healthcare plans offered by businesses, or Medicare and Medicaid plans. These teams tend to be highly specialized, and they focus on creating strategic approaches to content geared to a particular business need. But no matter where that team sits within an organization—and even if content strategy as a function is distributed throughout the organization—establishing a structure where content strategists can practice their trade goes a long way toward supporting the strength and longevity of the work of content strategy, or the core of the practice.

Also, departments and teams can be absorbed or completely dismantled. I've seen this happen where content strategists were reassigned to other types of content work, or worse, laid off or let go. I'm not saying that building a content strategy practice will safeguard you against those outcomes. But I am saying that *building a practice* with the support and buy-in of cross-functional teammates, product owners, and stakeholders *might* make the complete dismantling of the practice a less desirable option, especially after so many people have invested their time and resources into co-creating it with you, and especially because they have undoubtedly reaped the benefits of the practice as a result.

Beyond Copywriting: Meeting the Unmet Need

Imagine this scenario: You're the solo "content person" in your department or agency. You write copy for digital experiences, have a good understanding of UX principles, and you likely know a little bit about search engine optimization, or SEO.

You've heard of content strategy, but there's so much to learn. Then a client asks (and therefore makes the case) for the establishment of a content strategy practice, saying, "We hear that content strategy can help us create content that is performance-driven, useful, and reusable. Do you have anyone on your team who can do that for us?"

If you can relate to (or are currently experiencing) the previous scenario—or if you're a digital creative director, a content manager, or a user experience lead, and you've found yourself in a similar situation, take a deep breath, grab your favorite beverage, and settle into your favorite reading spot. There is ground to break and some structures to build. But first, you'll need to create and review the specs for getting it done.

POWER TOOLS RESOURCES ON THE *HOW* OF CONTENT STRATEGY

Since you're reading a book about building a content strategy practice, it's a safe bet that you've either done your research on, or know a thing or two about, what content strategy is, and you have a good idea of how it's done.

If, however, you're building a practice while simultaneously learning how to do content strategy, don't fret! Here's a short list of books to get you started:

- *The Web Content Strategist's Bible* by Richard Sheffield
- *Content Strategy for the Web* (2nd edition) by Kristina Halvorson and Melissa Rach
- *The Elements of Content Strategy* by Erin Kissane
- *Content Strategy at Work* by Margot Bloomstein
- *The Content Strategy Toolkit* by Meghan Casey
- *Writing Is Designing* by Michael J. Metts and Andy Welfle
- *Content Design* by Sarah Richards
- *Managing Enterprise Content* by Ann Rockley and Charles Cooper
- *Content Everywhere* by Sara Wachter-Boettcher

The first five books will dive into the "what and how" of content strategy. The next two track the evolution of content design from content strategy, and the last two take a deep dive into "back-end" content strategy—structuring content intelligently to make it future-ready and device agnostic. You'll learn more about front-end and back-end content strategy approaches in Chapter 4, "Expansion: Building Up or Building Out."

Blueprint Components

The second entry in Merriam Webster's online definition of a blueprint reads, "...something resembling a blueprint (as in serving as a model or providing guidance) especially: a detailed plan or program of action." It's that last part—a detailed plan or program of action—that parallels the concept of a blueprint as a tool for providing guidance as you consider the components necessary to build a content strategy practice.

There are five components to the practice building process that I've come to call the *Content Strategy Practice Blueprint*:

1. Making the business case
2. Building strong relationships with cross-functional teams
3. Creating frameworks and curating tools to build with
4. Rightsizing the practice to meet client or project demand
5. Establishing meaningful success measures

In this chapter, you'll learn how each blueprint component will help you build a practice that's both sustainable and scalable. In subsequent chapters, you'll learn about the people, procedures, and processes that support these components.

One last thing: Before you break ground, let's get aligned on the kind of content strategy practice that you're constructing. While this blueprint could apply to building a variety of practice types, our focus here is specifically about establishing *user experience-focused* content strategy practices—a practice that has the mission of creating and supporting a brand or organization's digital experiences and information spaces across digital channels, including websites and mobile apps, and that might extend to include AI, blockchain, and beyond.

Although the work of the practice may well include conducting content inventories and auditing content in social media spaces and on third-party websites, this book is not about content marketing strategy, which focuses on placing branded content (or content created in-house by a brand or organization) on third-party sites, social media, and similar channels.

Making the Business Case

In the building and construction trade, the circumstances that lead to breaking ground on a new building site are many, such as inheriting a new plot of land, or the need for more space, which necessitates acquiring adjacent plots to accommodate growth.

And so it is with building a content strategy practice.

Like a homeowner seeking a real-estate loan to make improvements that add value to a home, you'll want to show how building a content strategy practice adds value to your agency or organization. That's why the first component of the practice blueprint is making the business case. As well, every component that follows helps you implement this first step correctly and establish footing that is critical to creating a firm foundation for your practice as you build.

Conversely, there are other times when the business case is made *for* you. For example, there are creative leaders who realize that a client project—say, a website design—requires more than just a reskin and copy refresh. They know that something more deliberate and permanent is needed to support the sheer amount and types of content necessary to meet the needs of users and achieve the goals of the business, so they search for a content professional who can bring a critical skillset to complement an existing UX team.

Other times, there is a fierce advocate for content strategy of the user experience kind, who is willing to sponsor the establishment (or growth) of a practice that is distinct from marketing content operations—a practice that is focused on things like content structure, content hierarchy and the flow of information from one part of the experience to the next, and how things like navigational labels and visual cues help users find what they need and successfully complete tasks. That advocate may have hired a content strategist or two, or elevated an existing, seasoned, UX-leaning digital content pro to transition from content creation to content planning and other strategic functions to begin building out a practice.

Then there are situations where someone within an organization recognizes that adding content strategy to their user experience capabilities provides value to the business, where content is created and maintained as an asset. In this case, once a decision has been made to establish a team or practice, a UX or CX (customer experience) lead, manager, or director is tasked with staffing a content strategy team, and the people who comprise the team may eventually choose to formalize the practice.

No matter which of these scenarios you identify with, take the time to execute on the following steps to establish your footing and make the case for building your practice.

- **Know what you want to build before you break ground.** This book is about building a structure, or a practice, in order to house a function, which is known as *content strategy*. While it's true that there are overlaps between a team and a practice—and maybe you could argue that you can't build a practice without a team—you can, in fact, start a practice team of one and expand (or scale) that practice as the demand for content strategy increases.

- **Identify the value proposition that you'll share with business stakeholders.** This step involves communicating the value that the practice brings to your agency or organization, whether it is an expansion of agency capabilities and services you offer to your clients, or, for a midsized or enterprise practice, demonstrating how the practice can foster alignment around the strategic use of content to meet business goals and user needs.

- **Find relevant case studies—or create one from a past client or project.** Take this step to show how the establishment of a practice dedicated to delivering sustainable content strategies can make the difference for your clients—internal or external—by introducing repeatable processes for ensuring that content is useful and usable and supports the digital experiences created by your UX team.

So, whether you're lucky enough to have advocates clamoring for the creation of a content strategy practice that will create, curate and manage content as a vital asset to your agency or organization, or the business case is made for you, taking the time to walk through these three preliminary steps will help you avoid the risk of establishing your practice on an unstable foundation.

MAKING THE CASE FOR CONTENT STRATEGY

Barnali Banerji, Design and Research Manager, McAfee

When Barnali Banerji inherited a legacy team of UX writers (later called *content designers*) at McAfee, she knew there was a need to introduce content strategy into the mix. "The strategy part was essential because we have very complex apps and complex products that interconnect over different operating systems and different form factors. You need a content strategist who is able to see how to present content in an organized way—how to make content reusable, how to repurpose it, and how to establish consistency."

In order to differentiate between the types of content roles and the value each one could add, Banerji sought to better understand the role of the content designers on her team. "There was a lot of overlap with product design. So you start to ask, where does product design start? Where does it end? Where does content design start? And what is content design supposed to do?"

"I've done a lot of work on mental modeling and top task analysis, and how that reflects the information architecture of a product." The content designers were adept at storytelling as it related to products and services, but Banerji's team needed expertise in both areas—content design *and* content strategy."

Building Strong Relationships with Cross-Functional Teams

While the crafting of content may be a solo endeavor or one that's relegated to a team of writers, the effort it takes to bring that content to a screen or similar modality doesn't happen in a vacuum. There are many hands that your content must pass through before it becomes part of your digital experience, and the people (and disciplines) that those hands are attached to should be involved in the formation of your practice *before* you break ground. At the very least, these cross-functional disciplines should include the following:

- Visual designers
- User experience/human-centered designers

Banerji now has a mix of content disciplines on her team. "In my opinion, content strategy is very different from content design. When you look at strategy, you're actually talking about how might we present this offering, and how might we scale it? How might we measure that we're doing the right thing?"

Making the business case for (and showing the value of) content strategy at McAfee was easy to demonstrate for Banerji. "From the front-end perspective, because we have such complex apps, and because there are so many features on those apps which don't make sense to the user, that's where content strategy shows up."

The first content strategist to join Banerji's team helped to improve the information architecture. They showed how to organize information, and how to surface that information on the front end. The second strategist she hired helped to build a reusable and scalable content management system for their products.

Banerji has plans to add a third strategist to the team as well.

All of the content strategists are loosely embedded in product teams because Banerji wants them to have time for (and control over) being part of content governance discussions and driving more content-related initiatives at the broader level of the organization. "In order to do their best work, content strategists need to have a really good understanding of the business context and the larger product vision."

- User researchers
- Information architects
- Developers/engineers
- Product owners/managers
- Project managers

Introducing the concept of the practice and articulating its benefits, especially to the people you collaborate with the most, is a crucial step toward establishing practice longevity. At the early stages, you'll want to focus your efforts on understanding the functions or disciplines that are a part of your team; on helping every member of the team understand how content strategy will impact their work; and ultimately on helping everyone individually and collectively see the value that a strategic approach to content brings to your combined efforts.

You'll also want to consult with your colleagues as you begin the work of constructing a process framework. Involving them at this stage not only gets buy-in, but it also creates a sense of co-ownership in the practice. Chapter 2, "Structural Alignment," takes a closer look at how a content strategy practice can benefit each of the disciplines listed previously and provide you with some conversation starters on specific ways the practice can add value. You'll also learn what a process framework is and how to create one in Chapter 3.

Creating Frameworks and Curating Tools to Build With

To ensure stability and longevity, every structure, no matter how big or small requires a solid framework to help it stand. From the foundation to the footings, to the columns and beams and walls, every element that comprises a building's framework, along with the tools used to construct it, contributes to the strength of the structure, allowing it to withstand forces that would otherwise cause it to fail.

It's the same with the practice you're building. Creating a process framework—testing it and improving upon it—will help your practice stand strong. More than just a building metaphor, you will learn to create a repeatable framework that considers all the trades (cross-functional disciplines), tools, and elements that contribute to and are impacted by the work of your practice. You'll "soil test" your framework—meaning that you'll test the environment you're building in to ensure that you can create a firm foundation for your practice—with a variety of agency clients or in-house projects to show where you might need to add additional footings to further support the foundation of your practice, all with the goal of avoiding structural failure.

The following elements are critical to this blueprint component:

- The involvement of and collaboration with cross-functional teammates to establish alignment with the goals of the content strategy practice.
- The creation of an end-to-end process framework to identify responsibilities, dependencies, and critical handoffs in the development of a website or similar digital experience.
- The evaluation of a variety of tools to use within your practice at the project or client level to help you find what works best for your agency or organization.

Rightsizing the Practice to Meet Client or Project Demand

After you've successfully made the business case for the establishment of your agency or organization's content strategy practice, your next step is to rightsize the practice to fit demand. Rightsizing can sometimes have negative connotations, such as when an organization has to reduce its workforce to adjust to a downturn in business or market conditions. But in keeping with the building and construction theme, rightsizing in this instance refers to "creating a structure that's optimized for the size of the agency or organization—and for the number of clients or projects—where the practice is being built."

Even if there aren't any immediate projects on your radar, think bigger and consider the potential for future expansion of the structure you're currently building. This blueprint component requires you to think beyond the current project plans that are right in front of you and to consider how intentionally planning for future expansion can help you sustain practice growth. But how can you do that if the demand for the practice isn't there yet?

You create it. You use what you have in front of you to show how you'll grow the practice when the demand comes. Here's another way to approach it: most content strategy projects begin with a qualitative audit of the current state of the content and with a future state goal (usually informed by product requirements) in mind. If you are adhering to best practices, that future state goal usually includes plans to fill gaps in the content that might occur as a business or brand changes and grows.

You may not currently know what those changes will entail, but it's reasonable to assume that changes in your business goals will be constant. So you create a core content strategy that considers the content components you'll need to meet the current requirements, and one that also identifies content elements that may be needed in the future to support growth and change. And you'll also consider how that content can be structured for reuse across digital platforms.

Additionally, you should consider the people, processes, and tools needed, not only to maintain the core strategy, but also to accommodate change, including the estimated number and types of roles you'll need to fill, along with the workflows and governance needed to make key content decisions that will undoubtedly impact growth. Since you're likely already familiar with these processes from the *how* of content strategy, you can take a similar approach to rightsizing your practice.

If you are a sole practitioner at an agency or looking to rightsize your content operations into a more strategic position by expanding your practice to accommodate growing demand, consider adopting these approaches:

- **Are there current clients or projects you're involved with that will allow you to demonstrate the benefits of taking a more strategic approach to content?** If so, you can turn these projects into test cases by identifying a few quick wins that won't compromise the timeline as you demonstrate how the practice can scale to take on more work:

  - Take a proactive look at the organization's content through the lens of a sample inventory and audit (sometimes called a *spot audit*) and measure that content against attributes that map to future state goals.

  - Conduct a comparative analysis among similar brands or industry peers to identify potential content gaps that you can strategically turn into future content opportunities.

- **If there aren't any immediate client opportunities for you to work with, find out what's on tap with potential clients.** Determine if there is a chance for you to position content strategy as a value-add and to show how the practice can grow to accommodate more work:

  - If your agency is pitching its services on a rebranding project, ask to review the creative brief with an eye toward understanding why and how the potential client is planning to rebrand. Then do a quick spot audit of content on their website or app to see if current content offerings map to their future state, post-rebrand goals.

  - Determine if there are clients in need of an updated style guide or voice and tone guidelines. These are content deliverables that sometimes get overlooked and a demand that your practice can easily fulfill.

If you're a solo practitioner or the lone "content person" in a medium-sized organization, or if you're part of a small content team that wants to begin building a foundation for establishing a practice, the previous approaches can still work for you with a slight change of perspective:

- **First, you'll be looking at the content as an insider.** Hopefully, that means you have access to things like product backlogs and roadmaps to identify upcoming initiatives. You can take a proactive look at the types of content that may be needed to support those initiatives, and you can strategically turn the results of that review into content opportunities to demonstrate value.

- **Second, familiarity with your organization's content likely means you know a thing or two about the editorial process.** Conducting a few informal interviews with content creators, editors, SMEs (subject matter experts), and others involved with this process may reveal gaps or missed steps in that process that you can use to identify opportunities to improve the editorial process within the practice structure.

- **In many organizations, there's no single source of truth for information on who those content creators, editors, SMEs, and others are.** As a result, when the time comes for a site refresh or redesign, or migration to a new content management platform, the content team is left scrambling to figure out who owns what. This situation presents another great opportunity for you to demonstrate how the practice can meet another demand by creating a sample content matrix to show how a tool like this can save time and resources.

Establishing Shared and Meaningful Success Measures

There are several ways to measure the success of a singular content strategy, whether for an agency client or an in-house project. And there are just as many ways to determine whether the content created and curated based on that strategy is delivering against established metrics. Whether you use Key Performance Indicators (KPIs), Objectives and Key Results (OKRs), or some other metric du jour, your choices for measuring content effectiveness are many. There are also many ways for you to measure the success *of the practice*. Working with your client, content lead, or in-house digital experience team, along with product owners or managers and business stakeholders, can help you establish measures that are meaningful to the success of the practice and the overall success of your digital experience team.

Measuring effectiveness at both the project and practice level should:

- **Be collaborative**, involving colleagues and partners from multiple disciplines. If at an agency, this collaborative approach can (and should) be introduced to your client(s), and the results should be shared with those on your team who are responsible for delivering success for your client. Or your team can offer to lead a workshop to help clients establish this collaborative approach.

- **Be documented**, with the understanding that as business goals, user needs, and technologies change, so too will the things that you or your clients measure and how you measure them. You'll want to create a living document or repository of documents to track and update success measures as the practice matures.

- **Be shared**, not just as a finished artifact that gets circulated among teammates, or delivered to a client, and then shelved or archived never to again see the light of day, but to ensure buy-in and support from other departmental partners, as well as leadership for agency clients and in-house teams alike.

Ultimately, while there is no one single best measure of success for the practice, the measures you choose to gauge practice success depend largely on a few factors:

- **Agency-based or in-house practice:** Measures that matter to an agency practice may not apply (or may be unimportant) to an in-house practice within an organization. By the same token, the measures you ultimately decide upon to test the overall success of your practice will determine what measures make sense for your situation.

- **Overarching client or organizational goals:** While you can count the construction of your practice as a measure of success, your work doesn't end when the practice has been built. After the practice is standing and operational, project or client work will likely begin to increase in frequency and intensity as the word gets out that the practice is up and running. That means the success of the practice will be measured by meeting the goals that have been set by your client or dictated by the requirements of the project you're working on.

- **Quantifiable business goals:** These goals track the success of your practice for things that can be counted, usually as an answer to the question "How many?" as in "How many clients have we successfully provided content strategy services to?" or "Of the five most important content strategy initiatives we've identified, how many projects have we successfully completed?"

Keep in mind that an increase in the number of clients or projects is only a single factor among many to consider as you create and track success measures. You'll also want to look at whether projects were completed on time and to the satisfaction of the client contract or product brief.

- **Baseline number of clients or projects handled by the practice:** You may be tempted to assume that this number will always be zero, but if you've applied content strategy how-to to even a portion of a client or in-house project, those efforts count toward the establishment of growth measures, and will show where you are currently as you map out where you want to go in the practice's future state.

Renovations While Occupied

Chances are, if you work in content at an existing agency or organization, you're likely working on building a practice while simultaneously taking on your first content strategy project. While that might sound a bit like living in a house while an addition is being built, a.k.a. renovating while occupied, the truth is that in most cases, you simply can't avoid it. And that's OK. It's how you learn to build resiliency—and the strength and tenacity you create while doing so is how you'll eventually succeed.

This book is going to guide you as you build the structure *where* content strategy can happen—a solid container where the work can be done with minimal disruption. You'll learn the best ways to augment that structure so that the practice can function under different loads, whether at an agency with several clients, or within a medium organization or large enterprise.

You're also building a practice that invites stakeholders and teammates to co-work with you, both as trusted advisors and SMEs, and when it's time to hand off to another discipline in the product development process.

Establishing structure or building a practice gets buy-in at the earliest stages. If you decide to scale from a practice of one to a practice of many, you'll find that having that buy-in from stakeholders and teammates will make it easier to garner support from leadership as the practice grows.

TIPS FOR BUILDING A CONTENT TEAM

Andy Welfle, co-author, Writing Is Designing, *Head of Content Design, Adobe*

Andy Welfle knows a thing or two about building a UX content strategy practice from the ground up, having grown a team at Adobe from a solo operation to a team of ten, all *without* the benefit of a blueprint or guide. "I probably could have ramped up a lot faster if I had a book that talked about some of the common scenarios and things to look for."

Welfle had given little thought to organizational structure before joining Adobe. "I wasn't prepared for the open-endedness and ambiguity of everything." He soon figured out that work structure matters. "Who you report to and who your boss reports to sets you up for success—or failure."

Among the many books that have been written about organizational structure, Welfle wished there was one for content teams—especially teams that sit within a larger design org—and one that addresses typical organization structures and their strengths and weaknesses. "I was really lucky that I had a boss who let me figure things out and empowered me to say no to certain things."

Welfle was the lone content strategist among some 200 designers and 30 researchers. So knowing when to say no—and that it was OK to do so—was of particular importance to his personal well-being. Trying to be everything to everyone simply wasn't sustainable. "I definitely burnt myself out."

His boss suggested finding a product team to embed in and told Welfle that he wasn't expected to help everyone. "But I would get Slack messages from everyone looking for help. And I didn't want to say no. I wanted to show my value far and wide."

Getting product stakeholders to understand what he did and how his work added value took some work. "I tried, through a lot of trial and error, to get them to understand what I did and show my value as a content strategist. Some of it was hands-on explaining 'I'm going to help you through some of these content problems.' And some of it was just presenting a slide deck."

The need for establishing boundaries—and defining what kind of services were provided within those boundaries—became clear as the team grew. "Pretty early on we started doing office hours. It was really useful, not in solving actual problems, but for understanding the bigger problem space, to see trends and problems across products, and for building relationships. If I gave the perception of being accessible, people were a lot friendlier and willing to talk about that stuff. It worked out really well for me."

Welfle has spoken about his practice-building experience at conferences and meetups and has developed a list of six tips for growing a content team (see Figure 1.2). They're complementary to the practice blueprint outlined in this chapter, and they'll be useful to you as you're making the business case for building your content strategy practice.

6 tips for growing your content team

🙂 Get comfortable with ambiguity

☑ Learn the process (and show value)

💀 Talk about what you do. Constantly.

👥 Find your allies and build community

🧑 Be accessible (but set boundaries)

🏛 Care about org structure

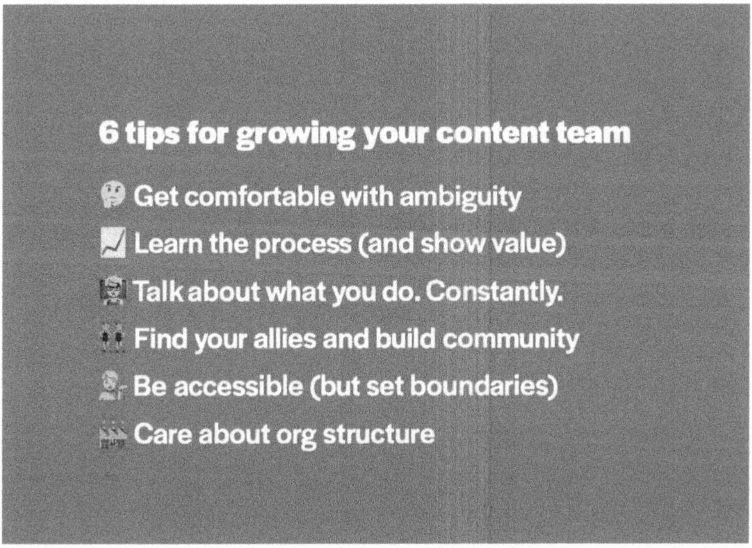

FIGURE 1.2
Andy Welfle's six tips for growing a content team are a useful addition to your toolbox and will be quite valuable as you are making the business case for building your content strategy practice.

A PRACTICE IN NEED OF A PLAN

 I've been herding digital content for nearly two decades, and even now, I'm still surprised at how many content strategists I've spoken with who can relate to starting out as the lone "content person" at some time in their career, evolving from solo web copywriters to growing or being a part of a team of content strategists.

Still, as this book started to come together, I wondered if my experience was more of an edge case. I mean, I knew how to make sense of smaller companies and agencies that lacked the resources to invest in building a content strategy practice. I could see how some didn't understand how establishing such a practice would level up their content creation and curation game—or that of their clients—while enhancing their organizations' user experience offerings.

And even though more and more books about the importance of the work and how to do it were being published with increasing frequency (in answer to an increasing industry demand), when it came time to stand up a practice—a figurative structure where the work gets done—there weren't many resources.

For example, when I first joined the agency where I stood up my first practice, I had to learn the hard way how to introduce the work to the staff UX lead. In my desire to do a good job, I inadvertently stepped on toes, and had to figure out, through trials and many, many errors, how to introduce content strategy—and by extension, the content strategy practice we'd eventually build together—to my cross-functional teammates.

I also had to learn how to articulate the value of the practice with everyone from designers to developers, to project managers and product owners, in a way that broke down potential barriers and built strong partnerships that helped the practice grow and thrive as an integral part of the agency's user experience capabilities.

But just because I was eventually able to get most of my colleagues on board, the work of building the practice was far from done. Like a builder with approved plans, I still had to figure out what materials I'd need to complete the practice-building process. I needed the frameworks and tools to create a firm foundation for building the practice, as well as success measures that I could point to as indicators of bringing value to the agency and its clients.

It wasn't until I moved to a larger organization with a similar practice-building goal that I began to document what worked (and what didn't) and to figure out if it could be scaled to fit a larger organization.

This book is the blueprint I wish I had access to years ago.

Persistent Principles to Remember Along the Journey

Wherever you are in your practice-building journey, you'll find it helpful to remember these principles along the way:

- You will always be called on to educate (or re-educate) clients, stakeholders, and team members about the value that the practice brings to your digital experience capabilities.

- You should always keep goals and success measures in mind as the scope of your projects—and the potential impact of the practice—evolves and changes.

- You'll find peace in knowing that you are not alone. There is a vibrant community of authors and experienced practitioners who have years of experience and lots and lots of stories to share.

The Punch List

There's a lot of information here for you to process, and there's even more actionable information to follow. For now, grasping these takeaways will prepare you for expanding your practice-building knowledge and skills, and set you up for success as you make your way through subsequent chapters:

- **There are only five components to the Content Strategy Practice Blueprint.** All five steps are integral to building a practice that can withstand stressors like tension and compression from outside of your practice structure. The order of the components (and the steps and guidelines within them) are important, too, but the main takeaway here is, even if you have to change the order of things, don't skip a component.

- **The work of content strategy isn't "one and done," nor is the construction of the practice where that work happens.** Just as a newly constructed building needs maintenance and repair to prevent structure failure, your practice will need similar attention to prevent its failure, too. Collaborating with cross-functional teammates, product owners and managers, and stakeholders to get buy-in and gain alignment will help you build resiliency and resistance to structure failure.

- **Remember the persistent principles:** Always educate, always highlight the practice goals, and always, always remember you are not alone. Your practice will always be stress tested, because things *will* get tough, no matter how hard you work to maintain your practice structure.

As an author and fellow practitioner, I've got you. And as a member of a generous community of passionate content strategy practitioners, we've got you. And you—yes you in the hard hat holding the blueprint and wondering what to do next—you've got this, too. Now let's learn the tools and tactics you're going to need to lay a strong foundation for your practice.

Structural Alignment

I *don't mean to be rude, but what exactly is it that you do?* You may be asked this question (or something similar) more than a few times in your content career, mostly by genuinely curious people who mean well, or by those who want to understand more about *content strategy* and specifically how it's different from *content marketing* strategy. There are other times when you may be asked this question by those who for some unknown reason have developed something akin to scorn for what they think content strategy is, and for how it impacts their work, or how they think it impacts their work.

In some instances, there may be a feeling that the discipline is redundant; that's where you'll hear comments like, *We already have content people* (or copywriters, or wordsmiths—you get the idea). In other instances, people feel that doing content strategy adds unnecessary delays to the product development process, as in *We already have issues getting the content when we need it. All this (extra) work is just going to slow us down even more.*

Don't be discouraged by these questions. Embrace and even encourage them. Because no matter the intent of the inquirer, it means that there are people around you who are curious enough to learn more, or who have strong opinions about content strategy as a discipline and are interested in the practice you're building. As you'll soon learn, the feedback you get when answering their questions will be quite useful in your ultimate goal of building alliances.

Defining Practice Space and Scope

At some point in your career, it might feel like your *only* job is to explain what content strategy *is*, what content strategists *do*, and why it's *important* to the product development process (not to mention its role in user needs and business goals). In fact, a good number of content strategy job descriptions seem to include a variation of this requirement: "Candidate must have the ability to clearly articulate the benefits of content strategy at all levels of the organization."

If you are the sole champion of content strategy in your agency or organization, the responsibility to articulate the vision and mission of the practice, as well as establishing clarity around the construction of the practice, falls directly on your shoulders. As you begin to socialize its benefits, you'll inevitably get asked a lot of questions, such as *Where does this practice fit organizationally speaking?* or *I'm a developer (or product manager or UX designer). What does content strategy have to do with me and the work that I do?* (Spoiler alert: Quite a bit, actually); or *How will the practice impact project timelines?* (Hint: Often, it's *not* what they think).

Remembering the first persistent principle from the previous chapter—that you will always be called upon to educate others about the value the practice brings to your digital experience capabilities—comes in handy here, as you'll find yourself explaining this often. The more you are asked to explain what content strategy is and what it does, the more you might find it's actually easier to begin by first explaining what content strategy *is not*, followed by an explanation of what it *is*. In broadening that explanation to include the concept of practice-building, this quote from an Agile training session comes to mind: By building a practice, you are "creating the boundaries within which the work is done." So, sometimes the best way to help people understand the parameters of the practice is to first establish what your practice *will not do* and what it *will not be* responsible for, which is just as important as what it *will do*. Although you are the subject matter expert (SME) on the topic, it's in your best interest to be as inclusive as possible when it comes to creating those boundaries, meaning that you should include as many of your cross-functional teammates as possible when deciding what your practice boundaries will be.

If inclusivity sounds counterintuitive to boundary-setting, remember this: the purpose of physical boundaries within a space, such as walls and doors, is more about defining the purpose of the space and what happens inside that space, and not always about blocking access to that space. For a real-world example, consider that while you can choose to enjoy a meal in your bedroom while binge-watching your favorite show on your laptop, that's not what the space was designed for, nor is it the intended use of the space. That is to say, generally bedrooms are for resting, and meals are usually had in an eat-in kitchen or dining room.

It's with that same idea in mind that you'll want to include your teammates and potential partners from other departments when defining the do's and don'ts of the practice you're building so that the purpose and intent is clearly defined to all who cross or encounter its figurative threshold.

In theory, it might seem like the list of things your practice will or won't do wouldn't be that difficult to develop. But when you take a closer look, you'll find that there are gray areas that need the input of other people or teams. Copywriters might be perplexed if your practitioners were responsible for creating content strategies *and* for implementing those same strategies by creating the copy as well. Or you may work with developers who wonder if URL and redirect strategies fall in your practice. Questions like these demonstrate why working with others to establish practice boundaries is key.

As you continue building, you'll also find that "one size fits all" does not apply to the creation of a content strategy practice. Nor does the work that's handled by the practice. The responsibilities of an agency-based practice primarily focused on client projects won't be the same as those handled by an enterprise practice, just like the do's and don'ts for a practice established within a healthcare organization won't work for a practice established within a consumer products company.

This is a really roundabout way of saying that the answer to the question, *What should the practice be responsible for?* is *It depends.*

The Importance of Alignment

For any structure to stand, the elements used to create it have what could be characterized as an agreement: each element has a job to do that contributes to the building's ability to stand and stay upright, based on the building's composition and function. When you establish a shared understanding around the purpose and importance of creating a solid content strategy practice with cross-functional teammates, or business unit partners from within your organization, or even with just one or two allies from other disciplines who share your vision, it works in a way that is similar to building components that work together to strengthen and uphold a structure. Except, unlike building elements, practice allies can talk and communicate, and ultimately agree on how to keep the practice standing.

This road will be lonely at first, especially if you are the sole champion for approaching content as a strategic asset for a client or your organization. Because you are so close to the content—the creation, maintenance, updating, and archival or removal of content that is no longer relevant—you hold the key to unlocking the potential that a strategic practice can bring to your agency or organization. It's a tough job, but with a little grit, a little sweat, and the right tools, it can be done.

But wouldn't it be easier to find some helpers—like the folks within your organization who don't have content in their title, or who aren't directly responsible for creating or curating content? Or the people who can see how the establishment of a content strategy practice could have a positive impact on everything from improving internal processes to the bottom line? You might think those people don't exist. But they do! They just might not know how to articulate their thoughts about content because they've never been asked. Or perhaps they aren't sure how to broach the subject, or who they should speak to about content stuff. And let's take it a step further and guess that there are those who don't yet know they feel a certain way about content. They may not have had their curiosity stirred by someone familiar enough with it to surface the importance of embracing the concept of content as a business asset, much less to mull over the idea of establishing a practice to handle the work of content strategy.

The key building component in both the previous scenarios is *you*: the content champion, the consensus builder, the content herder. You already know that content touches pretty much every aspect of a business or brand, both internally and externally. You also know that the consistency of content messaging is important, and that taking a holistic view of a brand, organization, or enterprise ensures consistency across all content touchpoints. That means you are in the *very best position* to create consensus across teams, departments, or business units so that everyone has a stake in the construction of the practice and benefitting from its success.

The venerable and wise Fred Rogers, host of the popular children's show, *Mister Rogers Neighborhood*, once said that as a child, when he saw things on the news that frightened him, his mother would tell him to look for the helpers—that he would always find people who were helping. So if the idea of establishing a shared understanding of the importance of building a content strategy practice seems daunting and scary to you, take a page from Mister Rogers' playbook and look for the helpers.

STRUCTURAL ALIGNMENT AND STRATEGIC PARTNERSHIPS

Jen Schmich, Senior Manager, UX Writing, Systems, and Infrastructure, Spotify

Jen Schmich built the first content strategy team inside product design at Intuit. She started by defining problems that writing would not solve, and she made sure to include stakeholders and other cross-functional teammates in those conversations.

Later in the process, she began sharing deliverables to demonstrate the difference between writing and content strategy (see Figure 2.1). As a content strategist who started out as a writer, she could speak to both disciplines from a place of deep knowledge. As she stated, "I'm showing examples of different deliverable sets: here's a content strategy; here's a content model. These are things that writers were not producing, and that no one had ever seen before."

Her team's success became evident as other departmental partners in the enterprise started vying to get content strategists on their projects. "Once they understood what we did and they valued our skills, people were saying, 'I've got to have some of that on my team.'" Although the content strategy team was initially focused on product design, over time they branched out into helping other teams and business units, including customer success, machine learning, and even data architecture.

In Schmich's experience, data architects are not language people. The data architects at Intuit recognized that fact, and they asked for content strategy to help them with the process of transforming technology into the semantic web. *What is semantic language? Who knows language?* Content strategy helped the data architects understand things like labels and nomenclature, e.g., *What do people call these things?* They also sought to understand how customers defined things, e.g., *What do these terms mean to customers?*

Data architecture is not where you'd expect to find content strategists, according to Schmich, "Partnering with them and looking at things from a completely different point of view was really valuable. We crossed boundaries we never thought we would have, but there is a big role to play in some of the lesser-known parts of the company that content strategy typically doesn't partner with."

CONTENT DESIGN	CONTENT STRATEGY
Making an experience through a design process to carry out a strategy	Forming content strategy to maximize the value of content creation
Emphasizes implementation of content	Plans along the entire content lifecycle
Focuses on instances of content	Focuses on abstract systems of content
Deliverables seen by users on a screen	Efforts are more often indirectly felt by users
A repeatable process	A distributed activity
Considers users its primary content consumer	Considers internal/external teams, machines, and users as consumers
A craft with competency in writing and design	A practice using broad content knowledge

COURTESY OF JENNIFER SCHMICH

FIGURE 2.1
Schmich created this visual to show the difference between content design and content strategy, and to foster alignment and understanding among stakeholders and other cross-functional teammates at Intuit.

Look for the allies and the content coalition-builders—those fierce advocates you learned about in the previous chapter who are eager to help you make the business case for building a practice. Look for those people who are in search of content practitioners who have the skills to augment a UX team by taking a strategic approach to things like labeling and nomenclature to help with wayfinding within a digital information space. Seek out those who can assess the current state of content and know what it will take to get to future state goals. When you start looking, you will see that the helpers have been there all along, just waiting for someone like you—someone with a holistic approach to all things content—to take up a shovel, break ground, and start building a practice that's focused on the importance of content and the value it brings to your customers, users, and your organization.

Building Alliances

The building of alliances supports the second component of the Content Strategy Practice Blueprint: building strong relationships with cross-functional teams —or playing well with others. This process includes taking the following steps:

- Understanding the roles and functions of each person on your team
- Asking your teammates what they know about content strategy
- Determining if the definition or understanding of content strategy that your team members and other partners have is different from yours and finding a way to bridge any gaps in understanding
- Providing as much information as necessary to foster alignment with how the practice is being structured
- Sharing how content strategy can benefit each discipline represented on the team
- Determining where the practice sits in the agency, organization, or enterprise

These steps are all part of the crucial footing you'll need to create to support the practice foundation and to ensure that your structure will stand.

The first part of the equation, that of understanding what your teammates bring to the table, along with what they understand about content strategy, might come naturally to you if you're already working in content. Research, fact finding, and interviewing, combined with having a good sense of who your audience is, are skills that most content creators rely on to craft content that is factually correct, findable, and usable. You can use those same skills to interview your teammates and to lay the groundwork for achieving alignment.

The second part of the equation calls for you to be able to share how content strategy complements other disciplines. This part might seem a bit more daunting. One way to start these conversations is to simply ask your teammates to identify pain points that they perceive to be related to content, and then to actively listen to what they share. Chances are some of what they share won't ultimately be the responsibility of content or the lack of a content strategy practice. But there are a few things at play here that you'll want to consider:

- Most people want to be heard and understood, especially about things that cause them stress or frustration when trying to do their best work.

- Taking the time to listen to and empathize with your teammates—actively listening without trying to interject your opinion or a solution—helps to establish trust.

- The list of pain points you glean from these conversations can actually become useful as you shift the conversation to sharing the benefits of establishing a content strategy practice. Here's a bonus—this list can also help you begin to formulate practice boundaries.

Another bonus that facilitating these conversations can yield is potential solutions for some of the frustrations that are not the responsibility of the practice to address, often because people simply aren't talking to each other. Your holistic view of the content, and by extension, the organization, can build bridges between teams and departments that can support healthy collaboration going forward.

Chapter 1, "The Content Strategy Practice Blueprint," contained a list of disciplines or roles that might be included on your digital experience team or within your agency's or organization's experience design department. These are the folks that you should definitely include when seeking to gain alignment around the structure of the practice, as all of these people (and the work they produce) will be impacted by the establishment of a content strategy practice in one way or another.

The list of disciplines or departments in your organization might be different, but the ultimate goal is the same:

- To understand roles and functions.
- To assess (or establish) a shared understanding of content strategy.
- To share how content strategy can benefit a function or department.
- To provide information needed to foster alignment with the structure of the practice.

Even though you may have established rapport with some of the people who represent these disciplines, you might find it hard to start up a conversation that has the ultimate goal of asking your teammates to support something that they don't yet understand. As you start doing exploratory research to gauge the level of knowledge your teammates have about content strategy, the results of that research will inform useful talking points that will provide context and understanding for your teammates.

In case you're not getting information that's useful for facilitating the conversations you need to have, or if you're still finding it hard to know how to start those conversations, here are some talking points on the benefits of content strategy as it relates to specific roles or disciplines that can help jump-start these critical discussions:

- **Visual designers:** Content strategy plans for the types and amount of content needed to support a digital experience. For a visual designer, that means establishing a process for having real (or as close to real as possible) content instead of placeholder copy (or the sometimes-dreaded *lorem ipsum*) *before* creating visual designs. This helps to avoid the frustrating and often deadline-busting hassle of having content "break" a design because the volume or type of content wasn't considered up front.
- **User experience or human-centered designers:** Content strategy and UX (or HCD) should be joined at the hip, focusing on the structure and flow of content from page to page, or screen to screen, as well as considering wayfinding and other aspects of the user experience. Establishing opportunities to co-work, as well as creating clear handoffs between disciplines, will yield a better product or experience for your client or organization's users or customers.

- **User researchers:** In addition to providing content for testing in user research sessions, content strategists sometimes partner with user researchers to contribute to conversation guides and research scripts for upcoming studies, as well as to co-create research readout reports.

- **Accessibility:** Content is foundational to bringing digital experiences to life—for *most* users. Whether working solo or with an accessibility team, content strategy helps to ensure equal access by removing barriers to digital experiences (for example, through the use of plain language). It also makes sure that experiences are inclusive and diverse by ensuring that the needs of all users are considered and supported through the quality of interaction that users have with your client's or organization's content.

- **Information architects:** In some organizations, content strategists also wear the hat of information architect (IA). It makes sense, given that both disciplines help users understand digital experiences and information spaces. Where both roles exist, IAs provide navigational tools that move users through an experience and make the experience seamless by the strategic placement of content throughout. Working together, content strategy and information architecture can achieve business goals and user needs by ensuring that users can find the information they need to complete a task.

- **Developers/engineers:** The content strategist's intimate knowledge of the content within a site or app goes well beyond the on-screen or front-end experience, and that knowledge can be quite useful to developers and engineers. That same strategist (or someone on their team) is likely familiar with how content is structured on the back end, through the creation of content models, or through the use of metadata and taxonomies that aid in content findability within the content management system (CMS). An established content strategy practice can also benefit engineering by standardizing approaches to site migrations, providing everything from target site URLs to redirects in support of the site development process.

- **Product managers:** The establishment of an intake process to gather future state content requirements provides a huge benefit to product professionals. That's because content strategists working within the practice can ensure that the content created in support of a product, service, or experience supports increased product adaptation and conversion, based on audience segmentation and other information provided by product managers. This partnership also facilitates on-time delivery of products and features, because content considerations are integrated *throughout* the product development process. This process helps avoid the last-minute dash for content that often happens just before a product is scheduled to launch.

- **Project managers:** Content strategists are intimately familiar with the level of effort involved in planning for, creating, and curating content that supports a digital experience. Project managers can benefit greatly from that familiarity and partner with content strategy practitioners to create processes that help keep project timelines on track. This partnership also ensures that important content-specific milestones and deliverables aren't overlooked.

If your agency or organization isn't structured in a way that involves the roles listed previously, you might be wondering if building alliances around the establishment of a content strategy practice is still possible, and if so, whom you should speak with about it? You'll be relieved to know that the establishment of a successful practice is not dependent upon the existence of that specific list. There are other people within your organization whom you can speak with to gain support and establish alliances as you build your practice, including the following:

- **Marketing:** In some organizations, content strategy *is* a marketing function. In others, it's solely UX. And if that's not confusing enough, sometimes content strategy sits in both marketing *and* user experience. Although this book is mainly about building a UX-focused content strategy practice, if that practice collaborates with marketing in any way, it's vitally important that you establish a shared understanding of content strategy with your marketing partners. Defining approaches, establishing swim lanes, and identifying any dependencies, such as content approval processes and handoffs that need to happen between your practice and the marketing team will strengthen the partnership as well.

- **Customer service/satisfaction:** There are organizations that have built entire content strategy teams around customer service, and specifically around the creation of a knowledge base. Whether that content is used as scripts for customer service agents or housed in an online experience for users and customers to access whenever help is needed, customer service and content strategy have a naturally symbiotic relationship. Who is better to glean customer pain points than the people on the front lines of customer service? And who can identify trends and turn those pain points into useful content to prevent those pain points better than a content strategist?

- **Technical support:** Similar to the relationship with customer service, content strategists and tech support are complementary in that content strategy can help translate technical jargon that's hard for users to understand into simpler language that makes an interface more intuitive. They can also identify patterns in technical issues to take back to the product team as recommendations for iterative improvements, serving as a bridge between these functions.

- **Other teams/departments:** Content strategists often collaborate with data analytics to identify patterns in contextualized historical data that informs content optimization opportunities for clients and in-house projects. You may also work with search engine optimization (SEO) to establish content strategy best practices that support content findability for both humans and bots.

As we shift the focus to project timelines and schedules, here's one last note for those working within an Agile organization or for those who have a program team that works on creating epics, or stories, that span more than a few sprints. If your team doesn't have a UX lead who represents all aspects of user experience inclusive of content strategy, it's vitally important to have someone from the practice either involved at the program level, or at the very least consulted by the program team whenever new projects and initiatives are being reviewed. This person ensures that the level of effort estimated for content strategy work is as close to accurate as possible, and makes sure that milestones, deliverables, and artifacts attributed to the practice (and individual projects) are correct. If content strategy isn't at the table when timelines are being established, there is a greater risk for the necessary time for all the phases to be shortchanged.

INCLUDING COLLEAGUES FOR CROSS-FUNCTIONAL ALIGNMENT

Candi Williams, author and Head of Content Design, Bumble

Working out how to embed the practice of content design in a large organization like Nationwide was a challenge for Candi Williams, but it was a really fun one. She started her tenure there as one of two content designers and eventually grew the team to 20. "I think people talk about scaling, and they think that headcount is going to be the answer to all of their problems when it's just the start of more problems."

As she grew the team at Nationwide, Williams focused on helping cross-functional colleagues understand why content practitioners did what they did, and she made sure to involve those colleagues in the process. She also made sure that people had a good understanding of the things they needed to know about the purpose of the practice and how it would help. "When you're in a big organization, no one actually really cares about what you do. They care about what you can do for them, and how you can work with them."

Showing the work of the practice also helped to foster alignment. For example, the team shared findings from a content audit that revealed duplicative content on mortgages across 50 pages by printing out the pages where that content was found, highlighting repetitive instances, and working through it with stakeholders. "Rather than saying 'Here's a spreadsheet. All of these pages have to go because they don't make sense,' we shared the work out and involved stakeholders in the process."

For Williams, alignment comes down to showing a keen and genuine interest in other people's roles and what they do—taking the time to understand their focus areas and their goals and identifying how you can work closely with them to achieve those goals. "It's about treating people like people and applying the focus that we usually apply to users and the people we're designing *for* on the people we're designing *with*—really getting to know them and their motivations."

Establishing Cadences and Aligning on Timelines

Content strategy sometimes gets a bad rap for "taking too long." In most, if not all cases, that bad rap happens for a few different reasons, most common among them—you guessed it—not including as many people as possible in the early stages of building alliances. While this refrain might sound like a broken record, it can't be stressed enough how vitally important it is to share the benefits of the practice with as many teammates and potential departmental partners as is possible *before* the practice is built and *before* the work of the practice gets started. This approach means that people won't feel like content strategy is something that is being *done to* them, or to their project, but instead that it is something that *includes* them, and has measurable benefits for their project. When people are included early and given a chance to ask questions to fully understand the practice concept and the work that will come out of it, there's usually less pushback.

Also, as you start to unpack the tools in your content strategy toolbox, and you begin using them to create strategies at the client or project level, the work you're doing might be overwhelming to those outside of the practice. When you do this work long enough, you'll likely come across tales of content strategists confidently presenting an approach that is research-based and thorough, only to have a product owner or manager interrupt to ask, "So exactly how long is all of this going to take?"

Or consider this story that a former colleague shared about a content strategist, who, in the middle of a presentation to a product owner was told by the PO, "I've seen enough"—but not in a way that suggested they'd been "sold" on what was being presented. In fact, after interrupting the presentation, the product owner ended the meeting, said a few terse words about the level of effort negatively impacting the timeline, and then abruptly walked out of the meeting room. That reaction was a bit extreme, and it's likely to be the exception and not the norm. Still, once you've completed your introductions and interviews, identified your allies, and thoroughly socialized what content strategy is, the value it brings, and why you are building a practice, it's vitally important to help those you partner with understand two very crucial things:

- **The work of the practice is scalable:** Not *every* approach, tool, or deliverable in your practice toolbox is needed for or applicable to every client or project.

- **The work of the practice is measurable:** Time invested in content strategy on the front end of a project alleviates the content scramble that often happens on the back end of that timeline, avoiding delays that can negatively impact ROI (return on investment).

In addition to these crucial points, there's one last thing about timelines that you'll want to share with your practice partners, and that is the idea of setting realistic expectations. That means letting everyone know that it *will* take time, trial, and yes—some unavoidable errors—to establish a cadence for the work performed within the practice as it settles and matures, and as clients and projects grow in size and scope. That also means that it's important for anyone outside of the practice who communicates practice timelines to confer with those directly involved with doing the work so that those timelines can be adjusted according to the scale of the work involved.

Communities of Practice and the Charters That Bind Them

Etienne and Beverly Wenger-Trayner are credited with creating the community of practice concept, which they define this way: "Communities of practice (CoPs) are groups of people who share a concern or a passion for something they do and learn how to do it better as they interact regularly."[1]

According to their website:

- The creation of a CoP allows for learning in what they call "a shared domain of interest."

- Members of a CoP form a community and build relationships that facilitate learning and knowledge sharing.

- Members are considered to be practitioners who share everything from tools to resources to help the creation and maintenance of a shared practice and develop a common way of sharing stories and case studies around their passion or interest.

---

1   Etienne and Beverly Wenger-Trayner, "Introduction to Communities of Practice," Wenger-Trayner.com, 2015, https://wenger-trayner.com/introduction-to-communities-of-practice/

UNREALISTIC EXPECTATIONS

There's nothing quite so fraught with mixed emotions than the realization that leadership has not only been sold on the importance of your content strategy practice, but they've also begun to broker conversations with potential clients or internal stakeholders about the potential benefits the practice can offer. Whether you've built your practice in an agency or across an enterprise, it's a great feeling to have your hard work both recognized and recommended.

On the one hand, there is the giddiness of having the work of the practice acknowledged. And on the other hand? There's the matter of the well-meaning but inaccurate representation of what the work encompasses and how long it takes to complete it, along with the sheer terror of learning that "a full content strategy" has been promised for a complicated website with over 5K pages and given a timeline that's more appropriate for a site that's a quarter of the size. Or that time when a truncated timeline was promised for a site migration that required stringent regulatory reviews based on the migration timeline for a similar migration that didn't require the same level of scrutiny.

Those important conversations at all levels of the organizations can help you avoid these mixed feelings, and more importantly, avoid underdelivering on a promise that sounded good, but wasn't based on reality. Establishing a practice charter, which you'll learn more about in the following section, puts the agreements you make about the purpose and operation of the practice in writing, along with how the practice approaches calculating project timelines.

One of the most important take-aways the CoP concept provided was the creation of a document called *The Community of Practice Charter*. According to the U.S.-based Centers for Disease Control and Prevention (CDC), which has developed a Communities of Practice resource kit to facilitate the creation of CoPs among public health professionals, "CoP charters, developed by each CoP, include mission, scope, objectives, and other course-setting components needed by the group."[2]

At a high level, the CDC's charter template covers the following:

- Introduction—Purpose of Community Charter
- Community Overview
- Justification
- Scope
- Community Participation
- Assumptions, Constraints, and Risks
- Community Organization
- Community Charter Approval

As you build a practice that is both sustainable and scalable, consider creating a practice charter to document the mission, purpose, and boundaries of the practice. This documentation serves these purposes:

- As a reference that can be passed along to new practitioners as they join your agency or organization as part of an onboarding strategy.

- As a guide for cross-functional teammates who are new to content strategy and who have a need to understand how the practice impacts their work.

- As an agreement to help course correct in the event there is a requirement or directive given that is outside the bounds of the agreed upon charter. There could be circumstances where you'll decide to modify the document or override the agreement to accommodate a specific client or stakeholder request, but there should be other processes in place to address changes in scope or responsibilities to support making such changes.

---

2   "Communities of Practice (CoPs)," Centers for Disease Control and Prevention, Public Health Professionals Gateway, last reviewed December 2, 2021, www.cdc.gov/phcommunities/index.html

If your organization isn't ready to build a formal content strategy practice, you can still establish a Community of Practice (CoP) as a place to identify the content strategy curious, as well as to provide a safe space for facilitating the types of conversations that will become invaluable once you are ready to build a more solid practice structure. There are many online resources available to help you with this endeavor, a few of which are listed here to help you get started:

- Scaled Agile Framework: Communities of Practice
  www.scaledagileframework.com/communities-of-practice/
- Stanford Communities of Practice: Content Strategy
  https://cop.stanford.edu/community/content-strategy
- "How to Build Your Content Strategy Community of Practice"
  www.braintraffic.com/articles/how-to-build-your-community-of-practice
- "Introduction to Communities of Practice"
  https://wenger-trayner.com/introduction-to-communities-of-practice/
- "Top 10 Tips to Create a Corporate Learning Community of Practice"
  https://elearningindustry.com/top-10-tips-create-corporate-learning-community-of-practice

Sustaining Alignment Through Shared Advocacy

You've come full circle to the main theme of this chapter: the importance of gaining alignment as a crucial first step in building a content strategy practice that's both sustainable and scalable. This step also sets expectations about the work to be done, shows how you'll collaborate, and establishes timelines for completing that work. You've also revisited the importance of the persistent principles outlined in the first chapter, specifically homing in on the first of the principles that addresses the frequency of being called upon to educate and reeducate others about the value of the practice. But the third principle—the peace of knowing that you are not alone, and that there is a community of authors and experienced practitioners with years of experience willing to share—is worthy of mention here as well, as you shift your focus to sustaining alignment and shared advocacy.

ESTABLISHING A CONTENT STRATEGY COP

 I once worked at an organization that had many interesting Slack channels available to join. There was one in particular that caught my eye: the Human-Centered Design Community of Practice (CoP). There was an entire channel created to host discussions about the integration of Human-Centered Design, or HCD, into the fabric of the company, as well as at the project level where appropriate. It was refreshing to see that there was a community of folks from across the organization representing many different disciplines who were passionate about HCD, and who had created a purposeful space to come together to share their passion, create a knowledge repository, and increase their knowledge on the topic.

I was intrigued by the concept of creating a *community* of practice, and curious to know how it was different from building a practice within an agency or organization, which, of course, is what this book is about.

The CoP concept worked well in an organization like this one, where there were several HCD practitioners, as well as those tangentially impacted by and therefore curious about it as an area of competency, and also where HCDs were somewhat siloed by projects. The HCD CoP was a great place to bring UX and HCD-related conundrums related to the specific project I was working on, and to get like-minded perspectives on how I might approach challenges through an HCD lens.

Leadership was receptive to learning more about content strategy as a practice, so I created a CoP dedicated to shared learning around the topic. Because of our project-specific structure there wasn't a demand for building out a full content strategy practice across the organization. But creating a content strategy CoP sowed seeds for the eventual creation of a practice if that demand should increase. In other words, the content strategy CoP identified those people who were curious about the discipline content strategy and provided a place for alliance-building to happen.

In formulating the content strategy CoP here, I drafted a practice charter as well. Although it was not as formal as the CDC's format, I included the following topics in the CoP charter (see Figure 2.2 and the bulleted list on page 46):

Content Strategy Community of Practice Charter

Introduction

"As content strategy continues to grow, the discipline changes all the time. One minute we're debating industry definitions, and the next minute we're tackling conversational design for kitchen appliances. The changing nature of our work makes it essential that we continue to [share with and] learn from each other."

—Tenessa Gemelke, *Brain Traffic* Blog[1]

Operating Definitions

- Content strategy guides the creation, delivery, and governance of useful, usable content.
- Content strategy means getting the right content to the right people, in the right place, at the right time.
- Content strategy is an integrated set of user-centered, goal-driven choices about content throughout its lifecycle.

—Kristina Halvorson, *Brain Traffic* Blog[2]

1 Tenessa Gemelke, "How to Teach Content Strategy," *Brain Traffic* (blog), July 19, 2018, www.braintraffic.com/insights/how-to-teach-content-strategy

2 Kristina Halvorson, "What Is Content Strategy? Connecting the Dots Between Disciplines," *Brain Traffic* (blog), October 26, 2017, www.braintraffic.com/insights/what-is-content-strategy

Value Proposition

Our Content Strategy CoP:

- Provides benefits from a shared knowledge and curiosity about content strategy as a complementary HCD + UX discipline.
- Contributes to a shared mastery of basic content strategy knowledge: how it fits within the company, as well as how it complements other practice areas.

Our content strategy CoP will:

1. Have a clear objective or purpose in mind. The purpose of the content strategy CoP is:
   - To share, teach, and grow knowledge about and competency in content strategy.
   - To encourage curiosity, facilitate conversation, and ask/answer questions about content strategy.
2. Establish (and continually co-create) our shared learning infrastructure. The initial focus of the content strategy CoP is to foster an exchange of information and knowledge sharing via
   - Slack channel for messaging and resource sharing
   - Ask Me Anything (AMAs), content clinics, lunch-and-learns, (cadence TBD)

3. Be aware of the experience and knowledge each team member brings to the table. Our content strategy CoP is for everybody. It is open to and enhanced by the participation of a diverse group of practitioners, including:
   - Content engineers
   - Designers
   - Developers
   - Information architects
   - UX writers
4. Hold an introductory meeting for interested members. Our first meeting will be hosted by the Human-Centered Design CoP:
   - [Date—month, day, time] and will include a "Content Strategy 101" presentation.
   - On demand: Presentation will be recorded and shared via Slack channel.
5. Have a moderator. This is key to the success of our CoP. For now, our CS CoP will be moderated by [facilitator]. As our shared knowledge grows, the moderator could be anyone with a passion for content strategy.
6. Hold regular meetings to share knowledge and strengthen communication. Meeting cadence TBD.
7. Use connectivity tools to keep in touch. Hello, Slack!

8. Use online tools/software that can serve as a virtual CoP headquarters. For example, we might use Trello to track things like discussion topics, resource lists, etc.
9. Offer support and resources to other CoPs. As the content strategy CoP grows, it will offer support and resources to other CoPs.
10. Hold occasional company-wide surveys to identify issues and areas of improvement that our CoP can help facilitate. These surveys will be used to:
    - Inform planning.
    - Determine issues or areas of weakness to focus on.
    - Focus on pressing matters (i.e., by project or across the organization).
    - Develop plans to move the practice forward.

FIGURE 2.2

The first draft of the Content Strategy Community of Practice Charter included agreed-upon definitions of what content strategy is, the value of the CoP to the company and clients, and a list of practice functions.

continues

- **Operating Definitions:** How we're defining (and practicing) content strategy
- **Value Proposition:**
    - What our CoP provides and contributes to the company
    - A specific list of things that include how the CoP is structured, along with meetings, activities, and other events to encourage engagement
- **Additional CoP Resources:** Links to articles and other repositories on forming and operating CoPs

You may notice that the draft charter didn't include a list of what our practice would *not* do. That was by design because:

- **Our CoP was a space for learning and doing.** Our CoP had the goal of encouraging participation and exploration into content strategy, and at this point, we hadn't been around long enough to discourage any topic or task for consideration.
- **Many CoPs have a lifecycle.** CoPs often have a clear beginning and end, where perhaps the group was organized to tackle a specific problem or objective, or when there is consensus on whether or not the value provided by the CoP has diminished. A CoP might end after solving whatever problem it was formed to address, or after it had met the stated objectives.

As membership in our content strategy CoP grew—and as we worked together to further refine what content strategy was and was not at this company, we allowed space for adding a list of "won't dos" to the charter. But as we were in the early stages of creating a community that was engaging and useful, all discussions and opinions were welcomed.

With few exceptions, many of the books written on content strategy all mention, in one way or another, the importance of shared advocacy. Part of the reason is that to some, content strategy is still new. And though it seems strange to think of this work as "new," all it takes is a cursory look around the web to see how much this work is needed, and how many brands still seem to be unaware of content strategy's many benefits.

The other thing is this: Where there is shared advocacy, there is also increased potential for sustaining alignment. Everyone who considers themselves a practice advocate has a stake in ensuring its success. Obviously, the first line of responsibility falls to you as the practice foreperson, along with other practitioners whose main goal is to create content strategies for your clients or stakeholders. But having the support of others who are even peripherally interested in practice sustainability means that there is buy-in across your agency or organization and a general consensus that the work of the practice is important and worthy of support.

Speaking in a webinar on leading content strategy across an enterprise, Anna Navarro-Schlegel, author, and Vice President of Global Portfolio Lifecycle Management at NetApp talked about the importance of speaking about and evangelizing content strategy. Navarro-Schlegel said she did two things to foster collaboration:

- She had lunch with as many people who wanted to attend to learn who they were, what they did, and what they knew.
- She identified members of leadership—those with manager or director titles—to start looking at the organization's content, and to ask, "What is it that you want [from this content]?"

As a result of this process Navarro-Schlegel said, " . . . you end up with a lot of clarity of what the organization needs. What they decided is that they wanted to lead the content strategy for the company. This is a tale of people collaborating and being stronger, giving them some power and ownership in the process. We have to find out who's passionate."

While this approach is specific to Navarro-Schlegel's company and circumstances, it parallels the alliance-building process outlined earlier in this chapter. Find the helpers—find the curious ones, the passionate ones, and even those initial dissenters—and you will find the components you need to create a firm foundation for your practice.[3]

---

3   Scott Abel et al., "Leading Content Strategy Across an Enterprise with Anna N Schlegel," *BrightTALK* (podcast), December 17, 2020, www.brighttalk.com/webcast/9273/425660/leading-content-strategy-across-an-enterprise-with-anna-n-schlegel

PEOPLE JUST WANT TO BE HEARD

Kristina Halvorson, founder and CEO of Brain Traffic, and author of Content Strategy for the Web

When it comes to seeking alignment around the practice of content strategy—or really any issue where there are differences of opinion—Kristina Halvorson says it all comes down to identifying and understanding core values. This is particularly true in organizations where there is discord between a UX-focused content strategy practice and marketing. "In any workplace culture, you have to pick your battles. There are going to be some marketers [who] are the loudest voices in the room. And sometimes you have to walk away."

But before walking away completely, you might be able to make some headway if you can identify and try to understand the core values of those you're trying to align with. What are the things that are important to them? What are the things their performance is being measured against? "It's understanding whatever goals they've been given by their supervisors in terms of what success looks like for them and having compassion for that as well."

Understanding what each team, department, or stakeholder (or your client) has been tasked with and then being able to reflect that back to them, along with understanding what others feel passionate about, can also help to foster alignment.

In the end, people—clients, cross-functional teammates, stakeholders, or departmental partners—just want to be heard. "If you can help make your project coworkers, team members, or people in other business units feel heard, that can get you far. It doesn't mean they're going to listen to you, but they're more likely to if they feel like you're listening to them."

The Punch List

Much of this stage of the practice building process focuses on people: establishing and building relationships and creating partnerships with those who have a vested interest in your practice. The inclusion of cross-functional team members and their divergent points of view at the beginning stages of the building process does the following:

- **Establishes a foundation** for co-creation and collaboration between content strategy practitioners and those from other disciplines.

- **Informs boundaries**—a list of dos and don'ts—for what will be within the purview of the practice and what will not.

- **Considers the impact of the practice** on existing project level timelines, and also paves the way for the inclusion of content strategy in those timelines, both at the practice and the project level.

- **Surfaces communication gaps** between disciplines that the practice can address (if appropriate) by building bridges, facilitating additional conversations, or by suggesting ways to shore up process gaps that result from lack of communication.

- **Creates space** to nurture a running dialogue to help sustain alignment through ongoing inclusion and advocacy.

While every component of the practice blueprint is important, the building of alliances through a little conversation and a lot of active listening is paramount for practice longevity and sustainability.

CHAPTER 3

Building Materials

About a decade ago, I was hired as a contract content strategist at an agency to work on a site redesign for a public health service client. I showed up on my first day with a laptop full of spreadsheets and templates, feeling confident for having done extensive client research, conducting an independent heuristic evaluation, and getting familiar with the public healthcare space. But as prepared as I thought I was, there were a few problems with my approach.

First, I parachuted into an existing digital experience team with an established website development process, and I failed to understand that process before jumping in with both feet.

Second, because I was the agency's first-ever content strategist, there seemed to be a lack of clarity around how—or where—my work fit into the team's existing processes.

After a few unproductive weeks of trying—and failing—to collaborate effectively with the UX lead, I set my spreadsheets and templates aside, and offered up a mea culpa. From there, the digital creative director, the UX lead, and I sequestered ourselves in a conference room lined with whiteboards to map out an end-to-end process—a process framework—to help assign responsibilities, create cooperation and collaboration among our disciplines, determine which discipline was responsible for producing deliverables, and identify critical handoffs at every point in the site development process. Without a process framework, it would have been nearly impossible to establish a foundation upon which to build a solid practice.

The Process Framework: Mapping from End-to-End

Just as a solid content strategy focuses on COPE—*create once, publish everywhere*—you'll want to establish a similar approach to building a sustainable practice, where you create a master framework that establishes an end-to-end process that can be used repeatedly.

COPE was first popularized by National Public Radio (NPR) as part of their digital editorial strategy. Their goal was to create content that could be easily adapted to and reused (or distributed) across devices, and across multiple channels, such as email, social media, text messages, and more.[1] The concept is used in content marketing as well, and also has relevance in UX-focused content strategy, especially as it relates to creating reusable content components and the practice of content modeling, where existing content—a product description, for example—is structured so that it can be used in multiple ways and across multiple channels.

In the context of practice-building, COPE represents the idea of creating an end-to-end process that is repeatable and scalable and is intentionally inclusive of your content strategy practice. This approach includes the practice (and practitioners) in the product or feature development process at the planning stage and is meant to ensure that content strategy is engaged throughout every stage of development, up to and including launch and beyond.

This last part—the beyond launch part—is vitally important to the longevity of your practice. That's because this approach to COPE is meant to demonstrate that the work of the practice is both reusable, as a process that can be used again and again, *and* repeatable, as a process that is replicable from project to project or from client to client.

Better still, reusability and repeatability speak to the efficiency of the practice you're building, and that efficiency can translate nicely to your agency or organization's bottom line. Here's a short list to help you get started.

- At an agency, as you bring your clients along the content strategy journey, you're not only teaching them how to think about and position their content strategically, but you're also introducing the concept of content strategy as a cyclical process, often referred to as the *content lifecycle* or *content strategy lifecycle*. Because this strategic approach to content is not "one and done," you have an opportunity to impact the possibility of repeat business with your clients.

---

1   Daniel Jacobson, "Clean Content = Portable Content," Inside npr.org, February 4, 2009, www.npr.org/sections/inside/2009/02/clean_content_portable_content.html

- At a mid-sized organization or enterprise, as you bring team-mates and stakeholders along in the content strategy journey, you'll be able to demonstrate how establishing a reusable and repeatable process can help your organization produce content more efficiently, and how to make sure that the content being produced continually tracks with business goals and user needs. From conducting inventories and audits, to creating a content matrix that tracks everything from who owns the content to when that content was last updated, to measuring content effectiveness and identifying optimization opportunities, these activities can save valuable time by avoiding the costly ineffi-ciency that a last-minute scramble for content can create.

NUTS AND BOLTS THE CONTENT LIFECYCLE

The five-phased content lifecycle was first introduced by content strategist Erin Scime, and later included in Meghan Casey's book, *The Content Strategy Toolkit*.[2] You'll find a diagram and more detailed discussion of the content lifecycle in Chapter 6, "Main-taining a Strong Core."

At a high level, Scime's lifecycle phases include analysis, strat-egy, plan, create, and maintain. When followed, these five phases ensure that digital content is not only findable, useful, and usable to users, but also that the production of that content—plan-ning, delivering, governing, and maintaining—is both sustainable and repeatable.

Similar to documenting customer or user journeys, or creating visual product roadmaps, you can take this same approach to map out an end-to-end process for your practice. You can use a physical white-board or a digital collaboration tool to map out phases by creating digital sticky notes and assigning different colors to different disci-plines, teams, or practices as shown in Figure 3.1. Then you can use symbols or icons to capture important content strategy milestones and critical handoffs between content strategy and other teams.

---

2   Meghan Casey, *The Content Strategy Toolkit* (San Francisco: New Riders, 2015), 203.

FIGURE 3.1
Using multicolored sticky notes, markers, and a whiteboard can be useful as you begin mapping out content strategy phases and listing sample deliverables that might be included in each phase.

Sample Phases of a Process Framework

You can create a Gantt chart like the one shown in Figure 3.2, which is based on the process framework document that we created at the agency where I'd inadvertently stepped on all those toes.

The following list is an example of project development phases you can use to begin creating your process framework. You can remove any phases that do not apply or add phases that are unique to your client or organization.

Research

- Review creative brief, product requirements, or Statement of Work
- Heuristic evaluation (existing digital experience)
- Content inventory
- Content audit
- Stakeholder interviews
- User interviews

Definition

- Personas (create, review, update)
- User goals
- User stories or scenarios
- Business requirements

Content Analysis

- Existing content analysis
- Gap analysis
- Comparative/peer analysis
- Editorial process review
- Readiness analysis (people and tech)

Site Definition and Structure

- Sitemap
- Card sorting
- Site framework definition

Design

- Design exploration
- Detailed designs
- Assets and audit (visual components)
- Prototype (if applicable)

Content Development

- Copy deck (for visual design and development teams)
- Content treatment (i.e., editorial, legal, or regulatory review, if applicable)
- Generate URLs if necessary

Project Management

- Backlog
- Deprioritized features
- Day 2 punch list items (fast follows)
- Maintenance phase

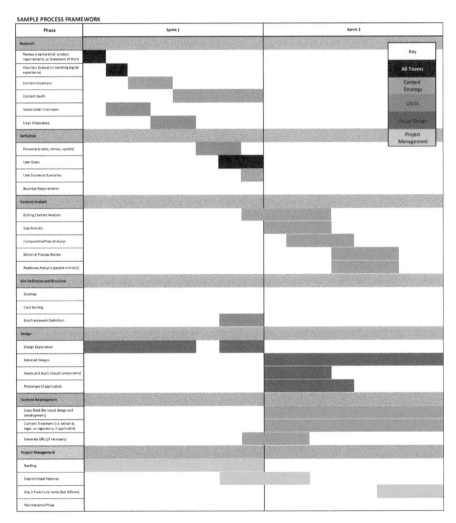

FIGURE 3.2

This simple visual of a process framework uses the sample phases from the previous list as a guide, and it is color coded to represent the discipline or team responsible for the work in each phase.

This chart represents the entirety of the creative website development process framework based on two-week sprints common to an iterative Agile approach. But if your agency or organization isn't an Agile shop, that's OK. Whether your organization follows an Agile or waterfall methodology, you'll want to be sure to capture the following items:

- Product development stages
- Duration of each stage
- Team(s) responsible for activities and deliverables
- Critical overlaps or handoffs between teams

At the start of the framework process, you'll want to enlist the help of some key players from each discipline within your experience design team—namely visual designers, UX leads, usability researchers, developers, and the like—to get buy-in, to be certain that every discipline involved in the digital development process is represented, and ultimately to ensure the success of the practice you're building.

NUTS AND BOLTS THE PROCESS FRAMEWORK AND
THE CONTENT LIFECYCLE

Just as the process framework follows the approaches taken to create or update a website or digital experience, the content lifecycle follows content from the analysis phase to development and implementation, and all the way through to content retirement, a.k.a. maintenance.

So, while the Gantt chart in Figure 3.2 shows the process framework as linear (or end-to end), the phases can also be connected in a circular fashion to represent a cyclical process.

Depending on the needs of your client or project, you can jump into the content lifecycle at any phase that is appropriate. And you can do the same with the process framework, shortening (or omitting) certain steps or phases that aren't needed or applicable to your project.

Framework Components and Considerations

The creation of a process framework might seem daunting, but if your agency or organization has a documented digital development process, the information in that process will go a long way toward helping you create the process framework for your practice that will become a part of (or at least inform) the overall development process. For practice building purposes, you'll want to focus on these areas:

- Process phases
- Artifacts and deliverables for each phase
- Disciplines responsible for phases, artifacts, and deliverables

The following list is a step-by-step approach for creating a process framework. You may find that some phases do not apply or that there are phases unique to your client or organization, such as submitting content for regulatory review, which are not represented here. The idea is to involve representatives from every discipline, from visual design to development, and to make sure that each critical step in the process is accounted for. From that view, you'll narrow your focus to user experience design, and then to content strategy, and map out your processes from end to end.

To create a process framework:

1. **Capture (in writing) as many phases** of your agency or organization's site development process as you can, from kickoff to launch, as well as any steps that are routinely taken post launch (sometimes referred to as *Day 2 items, fast follows*, or a *punch list*).

2. **Narrow your focus** on the phases of the process that user experience design is responsible for. Note that the level of granularity within each phase might change from client to client, or from project to project. You won't need to worry about those specifics just yet.

3. **Assign timeboxes for each phase**, capturing approximately how much time is needed to complete each one.

4. **Use colors, iconography, or other visual elements** to indicate which discipline is responsible for each phase. Note that these approximations may change as the scope of the project changes.

5. **Identify process overlaps.** For example, if your UXD team is small, you'll want to determine whether it's possible to collaborate for more efficiency—such as when conducting stakeholder or user interviews—or if a single discipline should be responsible for specific phases.

6. **Mind the gaps**—places where a phase or deliverable isn't accounted for but should be, such as generating a list of site URLs for development to inform site structure and to facilitate creating a demo environment.

7. Once your list of phases is complete, **take a second pass**, and have each discipline create a list of artifacts and deliverables that might be included in each phase. As you do so, you might find that you need to adjust timeboxes for completion of some phases.

Tools for Use Within Your Framework

If you are just getting started in the practice building trade, you'll be glad to know that there are some fantastic tools already in existence that you can test within your framework, and that you can use at the project level to help cement your practice. Remember that like the floors, columns, and beams of a building, these tools will support your framework, while helping your practice stand up and *stay* up, ultimately giving solo practitioners or a team of practitioners a solid structure in which to work. See the sidebar for some tools that are indispensable.

Before exploring these tools in more detail, a caveat: Some of the tools covered here will sound a lot like a "how-to guide," as in how to do content strategy. And while it is true that these resources do get specific about how to create a core strategy, the purpose of introducing them here is for you to know what's already out there and learn how to evaluate which tool (or combination of tools) is best suited for use in your practice, with specific clients, or on specific in-house projects.

The Web Content Strategist's Bible

The first of the two books I was given was Richard Sheffield's *The Web Content Strategist's Bible*. First published in 2009, Sheffield's book was written after he transitioned from tech writer to web content strategist, around the time web design and development were growing in popularity and scale.

Sheffield breaks down the website development process into seven phases:

- Proposal
- Discovery
- Analysis
- Design
- Build
- Test
- Maintenance

MY STARTER SET OF TOOLS

 When I began the work of standing up a content strategy practice for the first time, the only tools I had available were books that had been recommended to me by a product manager at another agency where I'd done some freelance work. Note that when I say *only*, it's not meant to diminish the importance of these tools. In fact, these tools continue to be foundational to my practice building—and practice sustaining—toolbox.

The job description for that first assignment was vague, if not completely confusing, so much so that I was a little nervous that the hiring manager had no idea what he needed. And I was even more worried that I wouldn't be able to figure it out.

Still, I showed up. And when I did, he gave me the URL of the client's current site, and the URL of the newly redesigned site, which was on staging. Then he handed me two books: *The Web Content Strategist's Bible* by Richard Sheffield and *Content Strategy for the Web* by Kristina Halvorson. Both volumes were heavily highlighted and dog-eared, marking pages that described the task at hand. With a sweep of his hand, he handed me the books and said, "This is what I need you to do." As it turned out, these books are part of a collection of tools that continue to be used within the practices I've helped to build, and you might find them indispensable to your practice as well.

These phases come in handy as a "how-to" for content strategists to follow when on a team responsible for building a new digital experience, for redesigning a new site or app, or for creating new features within an existing experience.

And when your focus is on building a practice, these phases, along with the tables, templates, and spreadsheets provided within, will help you show how this approach can be reused to deliver content that supports the digital experiences being created by your agency or in-house team.

Or, as Sheffield writes, you'll be able to show how "Content Strategy is a repeatable system that defines the entire editorial content development process for a website development project,"[3] and

---

3   Richard Sheffield, *The Web Content Strategist's Bible* (Atlanta, GA: CLUEFox Publishing, 2009), 35.

demonstrate how a "complete content strategy translates into more accurate contracts and more profitable [especially important for agency practitioners] and cost-effective projects [important for in-house practitioners]."[4]

Content Strategy for the Web

Kristina Halvorson and Melissa Rach's "little red book," *Content Strategy for the Web*, 2nd ed. is an integral tool for every content strategist, both as a how-to and as a foundational piece for building a practice.

The second edition includes information that's not only useful for individual content strategy projects, but also lends itself nicely to making the case for building a practice, including the following information:

- Expanded and restructured processes and tools for the research, development, and implementation phases of content strategy
- Recent case studies examining the impact content strategy has had on a variety of small and large organizations
- An examination of the ways that content-focused disciplines and job roles work together
- Discussion of the roadblocks you may encounter and the ways the field of content strategy continues to evolve[5]

The centerpiece of this book—the "Content Strategy Quad" (see Figure 3.3)—lends itself nicely to visualizing the phases of the work so that your leadership, stakeholders, and teammates, as well as any other partners you might collaborate with, can better understand how content strategy is done. It also provides another view of a tool that can be reused from project to project, or from client to client.

At the center of the quad is the core strategy—how content will accomplish business goals and meet user needs. From there, the authors break down the four components that together help achieve the core strategy, paraphrased here as follows:

- **Substance:** Identifying what content is needed to meet the needs of users and achieve business goals

---

4   Richard Sheffield, *The Web Content Strategist's Bible* (Atlanta, GA: CLUEFox Publishing, 2009), 36.

5   Kristina Halvorson and Melissa Rach, *Content Strategy for the Web*, 2nd ed. (Berkeley, CA: New Riders, 2012), xvii.

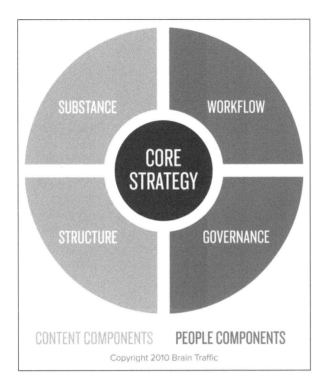

FIGURE 3.3
The original Content Strategy Quad, created by Kristina Halvorson, helps visualize the phases of content strategy work.

- **Structure:** Establishing how content will be prioritized and organized
- **Workflow:** Defining the tools and processes needed to establish and maintain the core strategy
- **Governance:** Documenting roles and responsibilities to determine who has oversight over key content decisions

Again, as with Sheffield's phases, the components in the quad represent reusable tools that you can use within your practice and will also help others re-envision content as a valuable business asset—one that will help organizations prioritize content initiatives (or those of your clients), streamline efforts, and foster an effective use of resources.[6]

---

6   Kristina Halvorson and Melissa Rach, *Content Strategy for the Web*, 2nd ed. (Berkeley, CA: New Riders, 2012), 27.

Enterprise Content Strategy: A Project Guide

As the title suggests, Kevin P. Nichols' *Enterprise Content Strategy: A Project Guide* is written for content strategists working at larger organizations or enterprises, although I'd argue that this work is applicable to content strategy work at organizations of any size.

I stumbled upon Nichols' slides from an Information Development World (IDW) presentation from October 2014. I'd moved from the agency to a larger organization as a Senior Content Strategist and was asked to find an approach to content strategy that could be used by the content strategists on our experience design team to help bridge a knowledge gap between us and our marketing partners. Specifically, we were seeking both an approach and an accompanying visual to show how our content strategy methodology was complementary to, yet distinct from, the work being done by content strategists on the marketing side of the house.

One slide in particular—slide 11—from Nichols' presentation seemed to meet our needs exactly. It spelled out visually and verbally a five-phase approach for getting the work done. Figure 3.4 is an adaptation of that slide and shows the five phases Nichols highlighted in his talk.

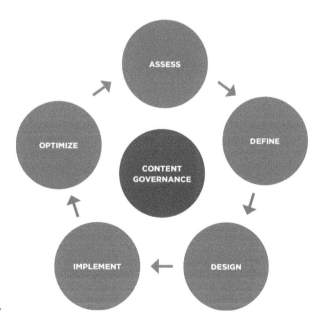

FIGURE 3.4
This adaptation of the five phases of content strategy work presented by Kevin P. Nichols represents another approach to getting the work done.

The following five phases that Nichols presented also fit in quite nicely with the Agile processes and ceremonies that our UX team adhered to:

- Assess
- Define
- Design
- Implement
- Optimize

We were able to use this approach to better facilitate conversations with our marketing partners. And we also used it to help strengthen and legitimize our practice, while educating stakeholders and cross-functional teammates about how we would include them at every phase of the process to ensure that we were delivering content strategies that met business goals and user needs.

Nichols' presentation and his five-phase approach were just the tip of the iceberg. When his book was published in 2015—a book that Nichols explained began "as a checklist of best practices for anyone interested in content strategy"[7]—it included an *eight*-phase approach he called the content strategy project lifecycle (see Figure 3.5[8]).

This closed-loop diagram centers around content governance. It strategically adds three additional phases to ensure that there's time built into the lifecycle to adequately plan before the assess phase begins; to account for the time it takes to build before publishing; and to highlight the importance of allowing additional time to measure content performance before taking additional steps to optimize it.

- Plan
- Assess
- Define
- Design
- Build
- Publish
- Measure
- Optimize

7   Kevin P. Nichols, *Enterprise Content Strategy* (Laguna Hills, CA: XML Press), vii.

8   Kevin P. Nichols, "The Next Generation of Content Strategy: Omnichannel, Performance-Driven Content, Content Marketing," October 25, 2014, www.slideshare.net/kpnichols/next-generationofcontentstrategyomnichannel perfomancedrivencontentkevinp-nichols

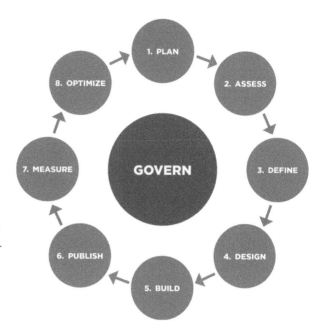

FIGURE 3.5
This figure is based on Kevin P. Nichols' eight-phased approach that comprises the content strategy project lifecycle.

At the project level, you may find that the inclusion of additional phases helps to visually outline how to take a more strategic approach of segmenting large content repositories, such as those found in higher education, financial, and healthcare spaces. Including these phases shows clients how content strategy is a repeatable process that lends itself to being flexible and agile enough to chunk and build a section at a time. At the practice level, having a multiphased approach in your toolbox—one that is adjustable and scalable—demonstrates to stakeholders the value that a dedicated practice can bring to clients and projects of all sizes.

Testing Your Framework: Tension and Compression

In the same way that the framework of a building—the floors, columns, and beams—are needed to ensure that a building stands up, your content strategy practice requires analogous elements to ensure that it stands up and stays up under the constant forces of tension and compression that it will inevitably have to withstand. Let's look at these concepts a little closer to understand how they relate to the building of your practice.

ALIGNED FOR SUCCESS

Back at the agency, it turned out that taking the time to sketch out an end-to-end process together with leads from visual design and UX helped to establish co-ownership, alignment, and a clear definition of responsibilities and handoffs between disciplines as we worked to redesign our public health client's website. There were some bumps along the way—and we had to pivot at the client's whim more than a few times. However, by the time the visual design and copy decks were ready to hand off to development, we had developed a solid process framework that helped us move from kickoff to launch with more clarity around milestones, and to establish ownership responsibilities for the activities within those milestones.

As that project was ending, I'd been told that there was a possibility of two additional clients interested in having us handle their web redesign projects, and who were both interested in hearing more about how content strategy might benefit their redesign projects. After conducting a sample inventory and audit of the content on each client's site and presenting our preliminary findings, we were hired to handle both projects. This gave us two more opportunities to test the strength and durability of our process framework.

As great as it was to get more work—and to have agency management believe in and sell the value of content strategy—it also became clear that to be successful, we not only needed to find another content strategist, but we also had an opportunity to establish content strategy as a core competency that we could offer to future clients.

Tension

In construction, tension happens when building materials are pulled or stretched. In the process of standing up an agency-based practice, you might encounter tension by being asked to take on tasks that are normally outside of the purview of content strategy—to be the wearer of multiple hats—instead of being able or allowed to focus on practice building. Or if you're focused on building an in-house practice at a mid-sized organization or enterprise, you might find yourself being pulled in different directions by the number and types of projects you're expected to manage before you've had a chance to establish your footing.

For example, if you work at an agency with a client (or multiple clients) moving to a new content management system (CMS) as part of a website update and redesign, you may be asked to train the client's in-house content team on how to use the CMS or develop a user guide for the users of the CMS, because you're "the content person." In many large companies, that task is often handled internally by content operations. In fact, it is best handled in-house, as there may be internal business processes or matters of security you may not be privy to. But if there isn't another person in your agency or on your team or in the client's organization to handle such requests, you may be asked to create training materials, causing the construction of your practice to grind to a halt.

Or maybe your organization's leadership has decided to ramp up the brand's social media presence, and because you "do content," you are asked to take on the work of content marketing—writing articles, social media posts, and related content marketing materials. After all, this work involves both "content" and "strategy," so it stands to reason that the responsibility would fall on your shoulders—at least to the people doing the asking. The argument could be made that these elements should rightfully be considered as part of a brand's overall content strategy, but they are not at the core of the UX-focused content strategy practice you are attempting to build.

This kind of tension pulls you away from the core of the practice—that of protecting the function of content strategy so that you can deliver on the promises of the practice that you hammered out with your cross-functional teammates. The creation of a strategic approach to content at every phase of a digital experience project that goes beyond copywriting balances the goals of the business with the needs of the user and supports the overall digital experience you're creating with your team.

If you've done the work of introducing the practice outlined in the previous chapter, then fear not. While this kind of tension isn't completely avoidable, you'll soon learn how to use these one-off requests and distractions to educate those outside of your immediate cross-functional team to show how what they're asking is distinct and separate from the type of practice you're building.

And if need be, you can also show them how distracting you from that purpose can be potentially costly. The work of the practice is to ensure that the content strategist has the space and structure to create strategies for the creation of content that clearly delivers the brand's value proposition, converting curious users to committed customers. Not doing so has the potential to negatively impact the return on investment (ROI), where the client or organization has invested in creating a digital experience—a website or app, for example. Distractions like the ones shared above can result in compromising the time needed to do the work thoroughly. As a result, important steps in the process, such as understanding the problem space, or conducting an inventory and audit, get skipped.

And there's also the matter of how a practice dedicated to the discipline of content strategy helps save money by contributing to estimating and scoping projects more accurately: by avoiding those dreaded last-minute moments where seemingly perfect project plans get derailed because content was left as an afterthought, or when existing copy disrupts the visual design or compromises functionality and needs to be re-edited.

Compression

Compression happens when building materials are pushed against or squeezed. As you're building your practice, compression may present itself as pushback from departments outside of your immediate cross-functional team.

For example, in many companies, UX-focused content strategy is co-located with visual designers, information architects, and other UX professionals. And in an Agile setting, UX practitioners may well be part of a larger team—a development, product, or scrum team—that is responsible for producing the work to support new products or features.

While that team may include UI/UX writers to create copy, chances are you'll be partnered with other departments to create copy, like marketing, where there's often a well-established editorial process that encompasses multiple communication channels, both online and off. If the work of marketing overlaps with (or is at odds with) the work of UX, you may get pushback that can stress the structure you're trying to create.

No matter whether you're experiencing tension or compression—or both—you'll want to focus your efforts on creating a process framework that not only ensures the longevity of the practice, but also addresses what the practice is, what it isn't, and how its function is different from other teams and departments. Doing so will ensure that your structure can withstand the stressors that are bound to test it, while allowing the practice to function as intended.

As you try out your framework and tools on different clients or projects, it may feel as though you are merely putting a process in place that is solely focused on the project level. And that's completely understandable. The thing to remember is that, as it is with a building that is under construction—where the framing for a single space such as a single room or office is only part of what will become a complete structure—you must start somewhere. The most logical starting place is with the work that's right in front of you, be it a single client or an in-house project. Remember the bigger picture you have in mind—to scale and grow what you've started, by creating the scaffolding and framing that will eventually become a solid structure, which is your practice.

So, when you've hit all the milestones of a project, and have tested and improved upon your framework to the point that it can withstand tension and compression, you're ready to test the process again with another client or another project, so that you can work out any rough spots that may need further improvement as you solidify the practice structure.

Stay Streamlined or Grow to Fit Demand

OK, your framing is up, and your work is taking shape. You're working within a structured practice, refining, and adding details where necessary. Like furnishings in a newly constructed space, you're moving things around to get the right layout, which for your practice may look like shifting milestones, or adding or subtracting deliverables based on the size and scope of the client's needs, or the needs of your organization.

Now what?

You may choose to keep your practice lean and streamlined, working with cross-functional colleagues to maintain the practice structure while proactively identifying opportunities to work with new digital clients who could benefit from the strategic approach to content that you as a practitioner can provide. At some point, there may be multiple clients that need the services your practice can provide. In that case, if you're a solo practitioner, you can approach the account manager and make the case for bringing on another content strategy practitioner dedicated to additional client projects. You've built a solid structure to accommodate such growth, and ideally there is room for you to bring another practitioner in, say as a temporary tenant (a contract practitioner) within the space you've created, who will occupy space in your structure for a time, or as a permanent tenant (a full-time practitioner), who will be on hand to take the overflow of work that comes your way. The same opportunities and choices are yours if you are working within an organization or large enterprise. If the work is there, then you'll need to grow the practice to accommodate the additional load. In either case, whether you stay lean and streamlined or grow to fit the demand, be sure to keep your teammates from other disciplines involved along the way so that they grow and scale with you.

NOTES FROM THE JOB SITE

PASSING INSPECTION

 The opportunity to test our agency's new process framework with two new clients yielded some interesting (and unexpected) outcomes. In both instances, the inclusion of content strategy in both client proposals signaled a win for being seen as "an added value" that the agency could offer to clients, which was great.

But there were problems as well. In one case, scope creep and a lack of understanding of the time needed to complete each phase of the content strategy work made it nearly impossible to keep up with the sprawl of the client's digital landscape. And while we were presented with an opportunity to improve on how to best represent and communicate timelines to clients, that project met with an untimely demise.

In the other case, project management miscommunicated the scope of content strategy work to be included under the client's existing contract and budget. Even though we may have taken a bit of a loss on that project, we ultimately received an Addy Award for our work with that client.

UNDER PRESSURE—A CASE STUDY

 After my initial but brief exposure to content strategy by that cadre of contractors sent to the online directory company where I worked early in my career, I was inspired to begin thinking of content more strategically. Although if I'm being honest, "inspired" is putting it mildly. The truth of the matter is that I was forced. Our online directory home page was prominently branded with our trademark colors and tagline. In addition to a traditional navigation structure across the top of the page, it also included a series of primary and secondary tabs that featured links to popular city pages, search terms, and categories, as well as a tertiary tab that was akin to a grab bag of tools and links to pages on the site that we wanted to drive traffic to.

Our parent company came to us with a challenge: find space on the home page to promote a contest. Not just any contest, but a huge promotion with prizes that included travel to high profile international events being sponsored in part by the parent company, which had made a sizeable investment. In other words, we needed to do our part to promote the heck out of this thing. The problem we faced was, just like our technology, our home page was inflexible. So much so that when there was a need to push time-sensitive content to the website, we had to ask development to do an emergency release.

But I'm getting ahead of myself here. Before we could deal with the limitations of our tech dependencies, we needed to figure out exactly where on the home page we could put any kind of content promos. And, knowing that we were required to produce multiple assets, we also had to figure out how to publish promotional content in a timely fashion without requiring an emergency release every single time.

Our marketing partners were the project owners of this endeavor, but our consumer experience team was responsible for architecting a solution. On top of that, there were other third-party companies involved in some of the promos, and a governing committee with stringent visual, voice, and tone requirements that we had to follow as well.

Did I mention that the digital solution had to appear on the home page? And that we also needed to create a landing page where users could find more detailed information? Talk about an information architecture nightmare.

Hello navigational dead-end? Meet the "Back" button!

As I began to digest not only the requirements, but also the type and amount of content we needed to accommodate, I wondered if it was time to admit I was in over my head. Still, I sat and stared at that home page for hours. I brainstormed, sketched, and workshopped with our marketing partners, visual designers, and development lead. And just when we were all ready to throw in the proverbial towel, a solution emerged: promotional modules!

The truth of the matter was, while it's usually desirable to have some visual breathing room within any visual design, there was so much wasted space on our home page that with a lot of thought and even more sketching, wireframing, and lo-fidelity designing, we came up with a solution that met the contest requirements by creating an area where we could place promotional "tiles" that were interchangeable and available in two sizes: one large (wider) tile, or two smaller sized tiles that could run side by side.

Our solution also provided us with a way to promote other products and features—like our early mobile apps—and it introduced the potential for monetizing those tiles with paid ads.

Then came the problem of workflow and governance—although those terms were still a bit foreign to me at the time—and the need to create some structure around things like:

- Requesting/reserving promo space
- Documenting promo requirements, both visual and text (and later, functionality)
- Managing the space by keeping an inventory of past promo-tions (to avoid redundancy) and scheduling future promos
- Creating evergreen "in-house" promos that could be used in the event there wasn't anything time-sensitive to promote
- And more granularly speaking, making sure that every promo call to action had a destination—in most cases, a landing page—and determining where that page "lived" in the main site navigation

continues

And on it went. While we were very proud of our solution, it ended up opening a can of worms, because there was no framework in place for how to manage the chaos that ensued after the initial contest was over. As you might imagine, everybody from marketing to the parent company wanted a promo. Eventually, we came up with a structure for managing the workflow and governance of these promos. And technically speaking, a solution was built that allowed us to push content live independent of major code releases.

I had no idea at the time—approximately 2007—that I'd stumbled into the discipline that would later be a game changer for me career-wise. Our team as well had unwittingly stumbled upon a solution that would literally change the face (the home page) of our website, and by extension, the user experience, by introducing a more modular design with reusable components that transformed how we positioned content as a driver for increasing conversions.

As Kevin P. Nichols described in the preface of his book, *Enterprise Content Strategy: A Project Guide,* my colleagues and I were "connecting the dots between content as a deliverable (in its final state) and the strategy behind that deliverable."

The Punch List

If you're feeling a bit overwhelmed by all the minutiae covered in this chapter—the frameworks, the phases, the steps, activities, artifacts, and deliverables—take a breath, take a step back, and remember this:

- While content strategy as a discipline is not a one and done approach, the creation of a process framework is—mostly. You may have to adjust process phases along the way to accommodate the unique needs of a client or project, but for the most part, your framework won't change much, if at all.

- There will be trial and error as you test your process framework and as you test different tools to try out within your framework. The more you test your framework, the quicker you will learn what works and what doesn't.

- Every client and project presents an opportunity for you to demonstrate how the practice provides benefits as a discipline that is complementary to user experience, visual design, and development, strategically connecting dots to ensure that the goals of the business and needs of the users are being met.

Remember that the goal here is longevity. Testing the strength of your practice to withstand tension and compression and adjusting where necessary will go a long way toward ensuring durability and permanence.

Expansion: Building Up or Building Out

As with a well-constructed building or public space that relies on prominent signage and other visual cues to help people get around, a good content strategy supports wayfinding by suggesting the inclusion of clear feedback to help users know where they are within a digital experience. (Think: "You Are Here" signage in a mall, similar to the image shown in Figure 4.1.[1])

This concept is highlighted in Jakob Nielsen's first usability heuristic—visibility of system status—which, in addition to helping users understand where they are, also helps identify the next best action to take.

NN/g
NNGROUP.COM

FIGURE 4.1
Jakob Nielsen's first of ten heuristics, "Visibility of System Status," focuses on the importance of helping users understand where they are within a system or experience and what the next best action is to take, based on their location.

---

1   Jakob Nielsen, "10 Usability Heuristics for User Interface Design," Nielsen Norman Group, November 15, 2020, www.nngroup.com/articles/ten-usability-heuristics/

In that same vein, let's take a moment to establish where you are in the building process: Chapter 1, "The Content Strategy Practice Blueprint," introduced the five components that comprise the content strategy blueprint that, when taken together, will contribute to making the business case for building your practice. Chapter 2, "Structural Alignment," provided a deep dive into the importance of building alliances in support of making the business case for the practice, and for creating a continuous culture of learning and collaboration. In Chapter 3, "Building Materials," you learned how to create a process framework and discovered tools to use within that framework at the project level. As for where to go next? This chapter is going to help you figure out why, when, and how to grow your practice, with a focus on sustainability.

The first rule of successful growth and scaling: continue to include cross-functional teammates and departmental partners in the process. They can help to identify and assess friction points encountered in the early stages of the building process, both at the practice and project level, and work with you to find resolution before you scale up or out. As well, be sure to enlist the help of practice allies when checking your existing framework for any gaps, cracks, or fissures in the process framework that need maintenance. By including teammates, it helps you keep everyone informed of changes to the size and scope of the practice that could impact the function and operation of the entire team, as well as other departments you partner with.

NUTS AND BOLTS GROWING VS. SCALING

This chapter uses the terms *growth* (or *growing*) and *scale* (or *scaling*) quite a bit, and sometimes interchangeably. As the chapter title suggests, you're going to encounter the word *expansion* often as well. Still, it might be helpful to understand the basic meanings of growing and scaling as they are used here:

- *Growing* happens along a continuum, and it usually requires increased resources to increase profits.
- *Scaling* happens incrementally, where iterative changes impact a product or service positively in a way that increases revenue or some other measure of success.

Complicating matters even more, there are concepts that combine both terms, like *scalable growth*. You'll be happy to know that, in the interest of clarity, that concept is *not* explored here.

Why Expand?

There are myriad reasons why the owner of a home or commercial structure might want or need to modify or expand a structure: an additional bedroom or bathroom might be needed to accommodate a growing family, while a commercial property owner might need to accommodate a prime tenant's growing staff. In the case of rezoning, where a mix of tenant types, including residential, commercial, and retail, is allowed, a property owner must be able to adapt to that change to stay competitive with other properties.

So how do these examples translate to expanding a content strategy practice? The example of the growing family or staff could represent an increase in client or project demand. The rezoning example could symbolize the need for a core content strategy practice built around website development to expand to include mobile practitioners, conversational interfaces, or other devices. No matter the impetus, before you expand your practice, you'll need to understand first when and why to do it before you begin finding the solution or figuring out how to do it. And before all that? You'll need to conduct a structural assessment to determine if the current state of the practice can withstand expansion.

Solo, Mid-Sized, and Enterprise— Defined

As the title suggests, this book is meant to help you build practices of various sizes, from solo to scaled. But what might that look like in your organization? The fact is that there are nuances to every agency or company that prevent a specific practice from fitting squarely into one box or another. Table 4.1 gives you an idea of what each practice type could look like and where the lines might be blurred. Note that the numbers included here, as well as the descriptors used (solo, mid-size, and enterprise), are representative of what I've encountered in organizations where I've worked or provided consulting services.

TABLE 4.1 DEFINING PRACTICE TYPE BY THE NUMBERS

Practice Type	Solo	Mid-Sized	Enterprise
Number of Practitioners	1–2 (can temporarily expand to accommodate client demands)	3–5	5 or more practitioners, sometimes comprising smaller, independent "practice units"
Number of Business Units or Teams Served	If an agency, it's usually seated within a single business unit, i.e., digital experience. If it's within a small company, it could be part of a small UX team or an editorial unit in and of itself.	If an agency, it could be assigned to a specific client or brand project team. If an in-house practice within a mid-sized organization, it could be part of a UX team, an editorial team, or a hybrid UX/marketing team.	In-house is usually assigned to specific products or lines of business. It may be one of many smaller practice units with a singular focus, where all units come together under a "center of excellence" or "community of practice model."

Assessing for Structural Durability

In the building trade, there are several ways to modify a structure to accommodate change and growth: you can build up by adding a second story to your existing structure, or you can build out by expanding the structure's footprint. In every case, a structure has to be assessed—inspected and stress tested—to ensure that it can handle the additional load of an extra story or extra floors, or in the case of building out, ensuring that augmented structures will avoid failure as a result of additional construction that impacts the existing framework.

In much the same way, you'll assess the practice you've built to determine if it can meet the demand of taking on additional practitioners, clients, or projects. Applying these concepts to your practice will tell you if your structure is sturdy enough to accommodate expansion. Depending on whether or not you've built your practice within an agency, mid-size organization, or large enterprise, these concepts will give you enough information to decide what expansion could look like for your specific practice. See Table 4.2 for examples of what building up or building out might involve based on your practice type.

TABLE 4.2 BUILDING UP OR BUILDING OUT, BY PRACTICE TYPE

Practice Type	Solo	Mid-Sized	Enterprise
Building Up	Adding agency practitioners to a client's in-house content team for the duration of a contract to execute recommended strategy work.	Adding practitioners to specific business units, with the practice serving as the single source of truth for how the work gets done.	Adding (or augmenting existing) "practice units" within the enterprise to major lines of business. These units might execute the work with a different focus, i.e., content design and creation vs. content structure and creating content models, but all teams come together under a practice community or center of excellence model.
Building Out	Adding practitioners to accommodate growing client demand, to handle larger projects and to advise account teams on project feasibility and timelines.	Adding practitioners to accommodate in-house project demand, or (in an Agile environment) to embed within new scrum teams that are added to an Agile Release Train (ART, a.k.a. team of teams).	Adding practitioners to your specific practice unit to accommodate increased in-house project demand or to embed within focused teams, such as a team fully dedicated to mobile apps or AI projects.

There are many ways to approach testing your practice for structural durability to determine if you're ready to grow or scale and how best to go about it. But since your work is rooted in user experience, why not apply some tried and true UX methodologies to analyze your practice structure and to better understand the problem space? By using methods that your cross-functional teammates are comfortable with, you can foster participation and encourage collaboration across any silos that may still exist. Let's take a look at how these assessments could be conducted, first through the lens of customer journey mapping and then through the process of service design, and specifically, the creation of a service blueprint.

Using Customer Journey Mapping to Expand Practice Operations

A customer journey map is a visual tool that helps user experience teams understand the paths a user or customer takes when interacting with a brand or digital experience in order to successfully complete a task or reach a goal. The process is purely from the end user's point of view, and considers not only the path taken, but also observes how the user is feeling about the process along the way. Now imagine adapting this approach as a way to figure out the current state of your practice, with the end goal being an established expansion objective, such as adding a certain number of content strategy clients within a certain amount of time or increasing the number of projects handled by the practice.

You can use a few different alternatives as substitutes for the customer in the journey. Table 4.3 provides examples of approaches you can take.

TABLE 4.3 JOURNEY MAPPING THE PRACTICE EXPERIENCE

Who Is Taking the Journey?	What Can the Journey Teach You About Expanding the Practice?
The Practice Builder (Likely, You)	This approach takes an introspective look at practice construction from the ground up, and captures the paths taken to build up the practice to its current state. Note any points of confusion, friction, or resistance that you encountered, where those points originated, and how (or if) resolution was reached. Consult teammates and others for unbiased input. Then take what you learn to identify potential roadblocks to practice expansion and to surface the best way to approach it, whether it is adding practitioners, securing additional tools and technologies to add to practice capabilities, or figuring out how to increase the number of projects in the practice portfolio.

continues

TABLE 4.3 CONTINUED

Who Is Taking the Journey?	What Can the Journey Teach You About Expanding the Practice?
The Cross-Functional Teammate	This approach follows the journey of cross-functional teammates in the practice-building process, from inception to the current state. Similar to the previous journey, you'll work to identify points of abrasion or a lack of cohesion from the point of view of those whose work has been impacted by the construction of the practice. Again, you'll take what you learn and apply your findings to improve upon any discord or strife that may have been revealed, and to figure out whether adjustments to the process framework, modifying handoffs, or creating more efficient co-working opportunities will help decrease abrasion and create a smoother path for expansion.
The Departmental Partner	Similar to the approach taken with your cross-functional teammates, you'll look to departmental partners identified during structural alignment to give feedback on what it's been like to interface with the practice, from its inception to reality. Your goal is to identify conflicts or other tensions and determine how (or if) resolution has (or can) alleviate problems that may impede expansion. Look for pathways of understanding that you can explore with departmental partners to help them stay abreast of and better understand processes that support the mission of the practice and identify opportunities to build a bridge to encourage buy-in and support for expanding the practice.

While customer journey mapping provides valuable information for content strategists at the project level by unpacking your solo practice journey, or the journey of those who interact with the practice, the ultimate goal is for you to use an approach that is common to most user experience practitioners to encourage participation and to adapt the approach to help identify pathways to growing or scaling your practice.

Using a Service Blueprint Model to Expand Practice Operations

Where a journey map focuses on a customer's interactions with a brand or digital experience to complete a task or goal, a service blueprint gets more granular and inclusive by adding, among other things, backstage actions or behind-the-scenes activities performed by the brand organization that support the customer's actions. If this process sounds daunting, don't fret: the idea isn't for you to become an expert in service design or even to create a perfect service blueprint.

Rather, by following the steps to create a service blueprint, you can reveal touchpoints or handoffs between disciplines and departments that interact with your practice and may need strengthening before you expand. The book *Service Design: From Insight to Implementation* by Andy Polaine, Lavrans Løvlie, and Ben Reason is a great resource that will give you an in-depth look into this design methodology and guide you in the creation of a service blueprint.

In a *Harvard Business Review* article titled "Designing Services That Deliver,"[2] author G. Lynn Shostack wrote, "A service blueprint allows a company to explore all the issues inherent in creating or managing a service." According to Shostack, some of those issues include:

- Identifying processes
- Isolating fail points
- Establishing time frames
- Analyzing profitability

Shostack wrote that investing time and resources in service design activities, and specifically in creating a service blueprint, "encourages creativity, preemptive problem solving, and controlled implementation. It can reduce the potential for failure and enhance management's ability to think effectively about new services." Similarly, you can apply the service blueprint model and the issues Shostack highlighted to help grow or scale your practice size and operations, as shown in Table 4.4.

---

2   G. Lynn Shostack, "Designing Services That Deliver," *Harvard Business Review*, January 1984, https://hbr.org/1984/01/designing-services-that-deliver

TABLE 4.4 SERVICE BLUEPRINT FOR PRACTICE-BUILDING

Issues to Explore	What to Look For	Questions to Ask
Identifying Processes	Look at your process framework, along with the tools you use within the framework to identify potential gaps in the services your practice provides. This assessment can reveal missed touchpoints in the process that, once resolved, can clear a path to growth.	• Do practice processes include clear touchpoints at critical junctures in the digital product development process? • Are there other teams or departments that the practice impacts indirectly that have been (unintentionally) excluded from the process?
Isolating Fail Points	Canvass cross-functional colleagues and conduct project retrospectives to identify any fail points in the process. It might include missed handoffs between disciplines or overlooking important editorial processes that extend project timelines.	• Where can you strengthen your processes? • How can existing fail points be repaired? • Where is a potential for failure? • If potential fail points are identified, how can they be strengthened?
Establishing Time Frames	Check in with project managers and others who may be impacted by practice timelines. Look at past projects and check to see if time frames assigned to milestones within the practice framework need to be adjusted.	• Has the practice handled enough projects to establish a baseline for smaller efforts, as well as a reference point for scaling up to handle larger projects? • How can you establish a definition of "done" for different-sized efforts?
Analyzing Profitability	For agency-based practices, profitability might look like an increase in the number of contracts that include content strategy services. For in-house practices, consider consulting product owners responsible for their own budgets to extract profitability, which may be expressed as "improved efficiency" or "time saved on delayed launches."	• Agency practices: Is the rate allocated to practice work realistic for the effort involved? • What can comparative analysis (and a little internet sleuthing) tell you about your agency's rates for the services the practice provides? • In-house practices: How does your organization calculate ROI for UX (inclusive of content strategy) initiatives?

Paths to Expansion: People or Processes

Maybe you don't identify as a content strategist, but you've been tasked with finding the resources needed to create a content strategy practice. At some point in your research, you've probably read that there is still some ambiguity around what content strategy is, what it means, what the work includes, and who is responsible for doing the work (and for that matter, what those people are called). Or if you are a practicing content strategist, you've likely come across a number of job descriptions that sound a lot like content strategy, but have titles like content designer, content architect, content engineer, or UX writer. That's because as a discipline, content strategy has continued to evolve and expand since the early pioneers burst onto the scene with books, tools, and methodologies. Remember, when adding people and capabilities to your practice, it's important to understand how that evolution impacts the expansion of your practice when you contemplate what an ideal future state looks like.

As you expand and gather insights and feedback on the health and stability of your practice in its current state, it's worth your while to take a look at the variety of content practitioners who have come from the core discipline, and to consider if expanding practice capabilities is sustainable over time, if one of the ways you want to expand is by adding headcount.

As a reminder, this book isn't a how-to guide for learning content strategy. Nor is it meant to be a directory of job titles or the job descriptions that accompany those titles. Still, as you look toward expansion, at the very least, you'll want to be familiar with the variety of roles that fall within the realm of UX content strategy. Then you should consider whether expanding to include these roles will help you future-proof your practice by adding practitioners who specialize in related competencies.

Which End Is Up?

In 2016, Ann Rockley wrote an article titled "Why You Need Two Types of Content Strategists,"[3] which appeared on Content Marketing Institute's (CMI) website. Rockley's piece effectively (and visually) split content strategists into two groups: "front-end" and "back-end" (see Figure 4.2), with the former described as having a love for "the content and the customer experience," and the latter as having a love for "structure, scalability, and technology."

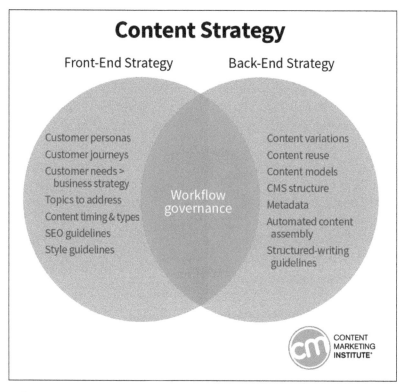

FIGURE 4.2

This Venn diagram is useful for visualizing the tasks of front-end and back-end content strategists, showing where they are different and where they overlap.[4]

---

3   Ann Rockley, "Why You Need Two Types of Content Strategists," Content Management Institute, February 22, 2016, https://contentmarketinginstitute.com/2016/02/types-content-strategist/

4   Ann Rockley, "Why You Need Two Types of Content Strategists," Content Management Institute, February 22, 2016, fig., https://contentmarketinginstitute.com/2016/02/types-content-strategist/

As CEO of The Rockley Group, and coauthor of one of the classic titles in the field, *Managing Enterprise Content: A Unified Content Strategy*, Rockley is known for building "intelligent structured content strategies," as well as working behind the scenes to build teams. She's also helped to firmly establish content strategy as a discipline. As such, Rockley knows a thing or two about content practitioners of all kinds, and she also knows a lot about growth. Given that many content strategists often need to move between front- and back-end skills, some people wonder whether or not the distinction is even necessary, or if such a (clear) divide even exists.

No matter what side of the diagram you fit into, in the context of a discussion about practice expansion, the most important things to focus on from Rockley's analysis are the tasks she's listed. This list can be used as a capability checklist and can give you a more granular look into the *what* of content strategy, giving you a point of reference for figuring out which of those capabilities, if any, you want to include as your practice expands.

A Content Strategist by Any Other Name

John Collins is a Senior Content Architect and Content Engineer at Atlassian. His title is not only uber-specific about what he does, but it also reflects some two decades of his work and experience in the world of content, and it clearly calls out his focus in that world. Collins authored a piece for LinkedIn Pulse[5] about content as a maturing discipline, chronicling his career path from tech writer to his current position, and including what Collins calls the "Four Emerging Roles in Content." Table 4.5 summarizes those roles and the responsibilities often assigned to them.

Note that this is just another way to define roles in content, which are ever-changing and evolving. The roles and responsibilities in your agency or organization might differ as well. Still this table provides another way to look at the myriad skills that content practitioners bring to the table.

---

5   John Collins, "The Maturing Content Discipline," LinkedIn, June 17, 2021, www.linkedin.com/pulse/maturing-content-discipline-john-collins/

TABLE 4.5 CONTENT ROLES AND RESPONSIBILITIES

Role	Responsibilities
Content Design	• Understands the user need. • Creates or curates content that meets user needs.
Content Strategy	• Considers how content can meet a user need. • Comes at the user need from a higher, more systematic level. • Applies a business lens to it at the same time.
Content Operations	• Organizes the shape, structure, and application of content (based on a definition by Cruce Saunders).
Content Engineering	• Gets content into a content management system (CMS) and then out to users. • Is more about people and processes. • Focuses on workflows, quality, governance, re-use, and so on.

Collins's four emerging roles and responsibilities (see Figure 4.3) provide another option for considering how to add content competencies as part of your expansion strategy. The roles and responsibilities in your agency or organization may be different—and some may not exist at all, at least not yet. Still, this list invites you to consider the following:

• Has your practice had to turn away or outsource work that could have been handled by one of the roles identified in Collin's list?

• Do you anticipate being involved with potential clients or in-house projects that would have a better chance of success if you had someone in one of the roles listed above?

Keep in mind that while in-house practitioners will have a better handle on future-state content opportunities, agency-based practitioners might not be able to anticipate this quite so readily. Taking a proactive look at long-standing clients will give you some information, but you'd need a crystal ball to help predict the needs of potential clients. Still, if there's a chance your agency wants to add practice capabilities, similar to Ann Rockley's Venn diagram in the previous section, this list or a similar resource can prepare you to have an informed discussion about what expanding capabilities might involve.

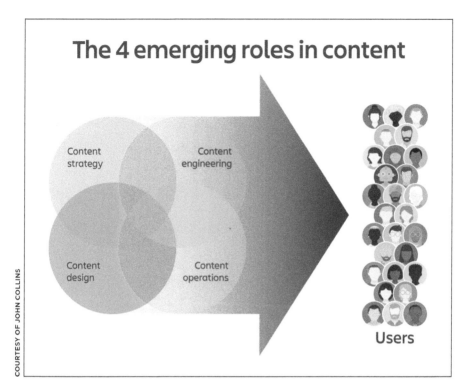

FIGURE 4.3

The four emerging roles in content, according to John Collins, provides another way to visualize the tasks and skillsets that may be included within a content strategy practice or content team.

Retooling Practice Processes to Scale

If budget constraints or other factors prevent you from adding more people resources to your practice, consider retooling your processes (see Chapter 7, "Retooling") to incrementally increase practice capability, with the caveat that practitioners should *never* be expected to take on more work than is reasonable or doable. But retooling does invite taking a different approach to demonstrating a path toward growth by establishing guidelines for scaling gradually in the event of an increase in demand for the services your practice provides.

In some cases, where adding full-time staff isn't feasible, contractors may be brought in to augment an agency or in-house practice. Having an established growth mitigation strategy also facilitates project orientation and onboarding, and it acts as another type of framework to further reinforce the structure of your practice.

Figure 4.4 shows one example of how to document a path to growth. It's based on a document created by a practice lead in a large organization and demonstrates how the in-house practice could scale, depending on the size and scope of a project. The phases of work listed here are based on the five-phase approach created by Kevin P. Nichols (introduced in Chapter 3), and they take the following factors into account:

- Content strategy approaches, methodologies, and best practices (i.e., inventories and audits, gap analysis, etc.)
- Phases of work, including how to assess, define, design, implement, and optimize
- Complexity, which varies by project (or client) type and by the scope of work
- Estimated number of pages (in this case, for a website) or URLs. These numbers can also be adapted to account for the end-to-end flow for mobile apps or other types of content assets.

Sample Growth Mitigation Strategy

	T-Shirt Size>>	XX-small	X-small	
Factors for Consideration	Story Points >>	1	2	
Approach, methodologies, best practices >		Consultation (e.g. simple content hierarchy or flow), or simple content audit to inform ideation process, or a one-off request.	Simple inventory/audit to identify content requirements to support small product / feature / functionality.	
Phases >	Description	Assess	Assess Define	
Complexity Factors >	Level of effort (LOE) based on brand impacts, multiple markets and/or states impacted/included in scope	Low	Low	
Estimated # of Pages or URLS in scope >	LOE for inventory, audit and other assessment work	1-5 pgs. or URLs	6-10 pgs. or URLs	

FIGURE 4.4

Here's one way to document a growth mitigation strategy for your practice, using T-shirt sizing—an Agile metric used for estimating and planning—to represent phases of work, complexity, and scope of work.

You'll notice that this table in Figure 4.4 includes both T-shirt sizing and story points as options for use within an Agile organization. If your practice isn't Agile based, then these sizing units can be ignored, or you can translate them into whatever estimation measures are used where your practice operates.

You also may have noticed that the practice in this scenario was not involved with implementation (i.e., writing copy) and is listed as consult-only for this phase. In this instance, there was a dedicated in-house team of UX writers (as well as writers on the marketing side of the house) to create copy. So if your practice uses this five-phased approach (or something similar) and you're responsible for the writing as well, you should modify your documentation accordingly.

In fact, everything shown in this table can and should be changed or modified in ways that are relevant to *your* practice. The main takeaway is to show how you can quite literally chart a path to growth by creating a mitigation strategy to demonstrate to others how the practice can achieve expansion goals.

Small	Medium	Large	X-Large
3	5	8	13
More exhaustive inventory, audit and analysis partnering with product and content partners to design future state experience.	Inventory, audit, analysis, recommendations to support future state feature. Incl. participating in/observing user research, competitive analysis, and consult on implementation.	Inventory, audit, analysis, recommendations to support future state feature. Incl. participating in/observing user research, stakeholder interviews/ presentations, competitive analysis, and consult on implementation, plus metadata, etc. Support for iterative product development.	Full application of five-phase CS framework, from kick-off to post - launch, to inc. recommendations to plan, create, and optimize content, including navigation categories and labels. Also includes metadata, taxonomy and related content structure info. Iterative support.
Assess Define Design	Assess Define Design Implementation (Consult Only)	Assess Define Design Implementation (Consult Only)	Assess Define Design Implementation (Consult Only) Optimize (post-launch test and assess)
Low to medium	Medium	High	High to very high
11-25 pgs. or URLs	26-75 pgs. or URLs	76-100 pgs. or URLs	101-500+ pgs. or URLs

GROWING PAINS

Meanwhile, back at the agency...

I started this book by sharing part of a story about how I'd been approached by a project manager who wanted to know where they could find more people like me. This happened after the content strategy deliverables I'd created and presented were (mostly) well received by a hard-to-please client. After seeing the value that the discipline (and practice) could add, the agency decided to offer the services of our fledgling practice of one—me—to another long-standing client. Their agreement meant more work for the practice—more than I could reasonably handle on my own—and that meant we needed to hire another me—stat!

Finding a strategist who could ramp up quickly and get started on an inventory and audit was hard. At that point in my career, I was still figuring out what to call myself, much less how to package and promote the work I was doing in a cohesive manner. I was a bit of a unicorn, and "finding another me" could only work if I first figured out who and what I was in the context of the agency and within our digital experience team. Still, the positive reception we'd received from that hard-to-please client definitely provided a boost to our collective confidence. But it didn't definitively answer the most important question, which was, "Can the practice scale?"

The answer? Yes...and.

Yes, because our UX team had worked together to create a repeatable and scalable process framework—and in support of that, I'd created a set of deliverable templates to be reused for future content strategy projects. Because we'd taken those steps, we felt we were in a good place to grow the practice to accommodate a second content strategy client and to bring in a second content strategist.

And, because we were willing to take a chance on a candidate with a content marketing background who showed a basic understanding of our UX-focused projects, not only were we able to take on that second client, but we were also able to proactively pitch our services to potential new clients as well.

Our second client was a utility company in the natural gas space with a unique challenge for us to solve: figure out how to create clear and engaging journeys for primary, secondary, and tertiary users, providing useful and actionable information for each user type while delivering on the message of the brand. Website visitors didn't know where to find information that was relevant to them, and nothing on the client's website screamed "convert [your utilities] to natural gas," which was a big deal for stakeholders.

Although the initial scope of work was focused on a visual rebranding and revamped copy, content strategy proved to be pivotal to the success of the project. Research with both users and stakeholders revealed that taking a magazine-like approach to depicting an aspirational aesthetic (referred to as "a natural gas lifestyle") was something homeowners could visually relate to. We provided that aesthetic while vastly improving wayfinding within the information spaces created on the site. This approach made a positive impact on conversions, and ultimately earned our agency an Addy Award.

But the story doesn't end there. We took on even more work. And at one point, we even expanded our capabilities by adding a few writers and editors to handle an overflow of work from other clients we'd started working with.

However, here is where I tell you that growth is hard. Winning over a difficult client felt good. And winning an award for some really good work felt awesome. But the third project we took on went up in flames, partially due to scope creep and the (unintentional) mismanagement of client expectations. The sheer amount of work needed to create a single cohesive experience out of an aggregate of multiple websites was beyond our capacity. We simply didn't have enough resources (or time) to do what was needed and to do it well. Still it was a learning experience that no one soon forgot, and it helped us discover (and sometimes bang our heads against) our practice boundaries.

I'll say it again: growth is hard. There are times you'll be rewarded for a job well done, and there will likely be many (many) more times that you'll make a misstep and have to regroup. And that's why they call it *growing pains*. That's where the growth really happens—in the midst of the mess.

The Punch List

There's more than one way to expand a practice: building up more often than not involves adding practitioners or headcount to your practice. Building out may also involve adding practitioners, and it also considers how refining and retooling existing processes and other resources can support incremental expansion. As you contemplate expanding your practice, keep the following steps in mind:

- Check the structural integrity of your practice to ensure that it can withstand the extra load that expansion will place upon it.
- Consult with cross-functional teammates and departmental partners to maintain alliances and garner feedback to use as additional inputs when assessing practice expansion plans.
- Stay aware of changes in the discipline that can inform new capabilities to add to the practice in order to meet future demands.
- If adding headcount isn't an option, chart a path to growth by proactively creating a mitigation plan to demonstrate how you can scale to accommodate projects with increased scope.

There is no single best way to expand your practice. Still, allowing change to happen organically (i.e., in response to increased demand), and charting a path that anticipates incremental growth, are the best options, instead of forcing expansion before you're ready. This type of planning and forecasting won't completely prevent the inevitable growing pains you'll experience as the practice matures, but hopefully it will lessen the intensity and increase practice resilience.

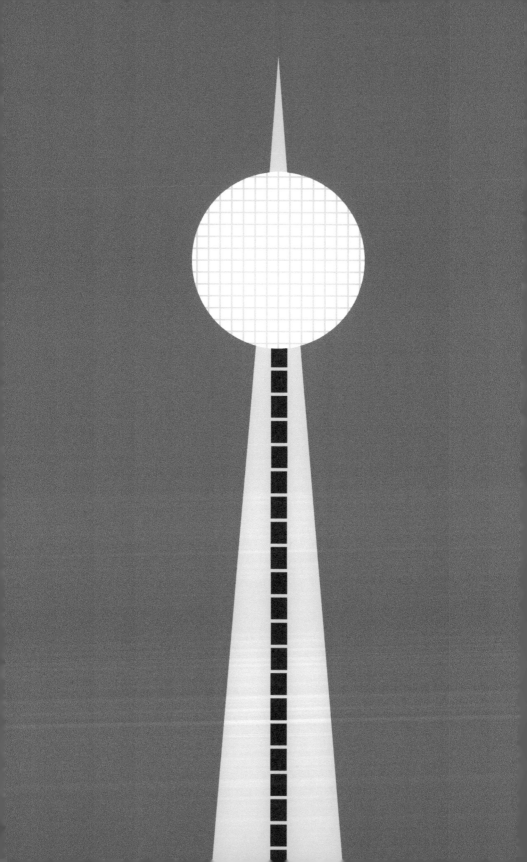

Measuring Success

The importance of measurement in the building and construction trade cannot be overstated. It can start with measuring the dimensions of a plot of land where a new structure will be built, or perhaps with the assessment of load-bearing walls and other components in an existing building, to make sure that it can withstand expansion. The significance of measurement in construction impacts time and materials costs, as well as projections for labor needed to complete a project. And, of course, the accurate measurement of the raw materials used to erect a structure is critical to ensure that the structure is built according to specifications, ultimately impacting safety, structural integrity, and longevity. Taken together, all of the ways that measurement is used and relied upon in building and construction provide a way for a construction company to assess its performance and gauge success.

Establishing meaningful content metrics on a *project-by-project* basis is an important step in the execution of a comprehensive content strategy as well. Metrics such as bounce rates, time on page, time on task, and task completion help measure performance, and data-driven content strategies are certainly important from a *how-to do* content strategy standpoint. But the focus of this chapter is about establishing and tracking *practice-level* measures and then achieving those objectives over a set period of time so that you have *quantifiable* data that you can point to as measurable indicators of how the practice continually adds value to your organization, your agency, or your clients.

So instead of measuring success with more traditional metrics, you'll learn to look for things that are more esoteric—and maybe even a bit more abstract—things that, at least initially, may not sound like vital measures to track, but that will act as gauges to measure the health and success of your practice in ways that are relevant to your business or organization.

Charting a Course to Success

There's an African proverb that says, "If you want to know the end, look at the beginning." In other words, to get an idea of where you're going to end up, consider where you were at the start—and consider everything that came in between—the wins, the losses, and especially the lessons learned along the journey.

For the purpose of this discussion, if you think of endings as goals you want to achieve as a result of executing a set of actions, then now is a good time to revisit some of the principles and processes covered earlier in this book—the beginning—before charting the course to the achievement of a successful content strategy practice—the end.

As you've read many times, although projects often focus on an end-to-end process that suggests a beginning and a hard stop at the end, in reality, for the work of content strategy (and your practice) to be counted as both valuable and successful, you'll come to the "end" of a project, and then rinse and repeat approaches and processes as you begin the cycle again. Your goal is to make sure that your client or organization's content stays relevant, avoids redundancy, and doesn't become outdated. That's the content lifecycle model in practice, and that lifecycle approach is key to establishing success measures.

Looking Back and Moving Forward

Using the wise words of that African proverb as a guide, let's walk back through a few concepts covered in earlier chapters and look ahead to future chapters, to understand how they can help you chart a course to success, with a focus on measurability:

- **The Process Framework:** Perhaps one of the most important documents created in the establishment of your practice, the process framework introduced in Chapter 3, "Building Materials," maps end-to-end practice processes. It also includes what work gets done and when, as well as who you will do the work with by identifying cross-functional teammates that you'll engage at specific milestones in the product development process. From the standpoint of charting practice success, the process framework is also a living document: as it is revised and updated, you may find new opportunities for the practice to partner across the organization. These new opportunities can be counted as practice success measures as well.

- **The Service Blueprint:** Chapter 4, "Expansion: Building Up or Building Out," also introduced the creation of a service blueprint as a way to reveal touchpoints or handoffs between the content strategy practice and other disciplines. These interactions are both measurable and scalable and can be included as measurements of success. For example, in the early establishment of the practice, your focus may have been solely on doing the nuts and bolts work of content strategy, such as conducting inventories and audits, as well as creating and curating content to meet business goals and user needs. As you scale practice operations, you may find opportunities to increase the number of departments within the business that you work with on projects or to create content strategy deliverables. All of this is measurable, too, and it all counts as you chart a path to success.

- **The Journey Map:** As discussed in Chapter 4, journey mapping is a technique used to establish future-state goals. Mapping the practice journey can uncover additional practice-level objectives. At an agency, you can count an increase in the number of clients the practice has worked with, or the number of clients who have turned into repeat customers, as a success measure. Within an organization, an increase in the number and types of projects completed counts toward practice success.

- **The Practice Playbook:** The practice playbook, which you'll read more about in Chapter 7, "Retooling," is a living document that includes practice processes and procedures, as well as a repository of practice artifacts, all of which are likely to change and grow over time as your organization grows, or as your agency takes on more clients. In and of itself, the creation of the playbook is a goal that counts as a measure of success. As a bonus, each time you make major revisions to the playbook—say, annually—it counts toward practice success as well, provided that you meet any deadlines you've established for doing so.

- **The Practice Roadmap:** The creation of a practice roadmap, which will also be covered in Chapter 7, provides a way for you to map how the practice can support your agency or organization. It's also a tool used to make decisions about the work the practice will do *outside of* product or content development. As you chart the steps to measurable success, the creation and maintenance of this roadmap invites you to innovate and ideate on work you can do and services you can provide beyond the day-to-day (or sprint-to-sprint) content work that is the foundation of your practice. The practice roadmap examples provided in Chapter 7—creating and maintaining a content matrix, creating templates and related documents to use with internal or external clients, building a content strategy knowledge base, and so on— all of these things are measurable, and as they are completed, they can be counted toward the overall success of the practice.

This approach to looking back at the beginning stages of practice building—the starting point or origin point—is meant to help you get crystal clear on what success looks like for your practice, and most relevant to this chapter, how you'll measure that success.

Before we continue, this caveat bears repeating: As with many other topics and techniques covered in this book, what success looks like for an agency-based practice will look different from what counts as success in a large organization or enterprise. The most important thing to grasp is that there are a variety of ways that you can gauge the success of your practice—and, as we'll cover later in this chapter—how you can tell if it's failing and what can be done about it. Take what's covered here and apply it to your particular situation so that you can track, measure, and if necessary, course correct as you move toward success.

Practice-Specific OKRs and Success Measures

OK, so you've got your process framework figured out. You know who your cross-functional partners are, where (and when) critical handoffs need to happen, and you're mining the framework for new partnership opportunities that you can measure. Maybe you've invited your UX design colleagues to participate in a journey mapping or service blueprint exercise to plant seeds for scaling the practice. And lo and behold, you now have a good idea of what success looks like. That's awesome! So how are you going to measure it? What additional ways can you quantify results that not only show that the practice has been successful, but that it's also ready to grow and scale? Here are a few more ideas you can use to discover additional success measures for your practice.

Office Hours

Content strategy practitioners love to solve content conundrums. We see things—patterns (or broken patterns), missing or confusing information hierarchies, or inconsistent labels and nomenclature within a single experience—that make us both anxious (to find) and excited (to fix). Depending on how the practice is structured, content strategists may not be embedded in every team or business unit across an organization. Keep in mind that at an agency, sometimes clients don't have the budget for content strategy, leaving visual designers to sort out content problems on their own.

MEASURING SUCCESS AT INTUIT

Jen Schmich, Senior Manager, UX Writing, Systems & Infrastructure, Spotify

In her role as a content design leader—and later, as senior manager of content systems at Intuit—Jen Schmich set out on a journey to build a content strategy team from the ground up.

She started by making the case for content strategy, explaining what other skill sets and perspectives existed within the content space by differentiating the work of content strategy from the work of writers. "We identified a whole set of problems that were content problems. And none of those problems would be solved by writers or by hiring more writers."

Next, Schmich identified a list of problems from the perspective of content strategy. She wrote about the team-building process on *Medium*[1] and shared that list. "Not one item on this list could be solved by adding writers."

- Content was inconsistent.
- Content updates took too long.
- Content couldn't be found.
- The wrong content could show up (in a product or experience).
- We created the same content over and over (redundancy).
- Tools created errors and wasted time.
- Things fell behind.

Schmich's team-building process also included working with stakeholders to align on problem areas. "Ultimately, you want to come out on the other side with partners and teams that are going to commit resources to fixing those problems."

---

1   Jennifer Schmich, "This Is How We Built a Content Strategy Team from the Ground Up," *UX Collective* (blog), *Medium*, September 3, 2019, https://uxdesign.cc/this-is-how-we-built-a-content-strategy-team-from-the-ground-up-91384dfd30d4

The measures of success for Intuit's content strategy team resulted from quantifying how the content strategy team grew in response to solving those problems, shown in Figure 5.1. "In the beginning, it was about multiplying partners and projects. We looked at the goals for the year overall and asked, what could we go for?"

TEAM GROWTH OVER TWO YEARS	
0 team members	5 team members
1 partner team	9 partner teams
1 business unit's scope	4 business unit's scope
0 work requests for our team a month	33 requests received a month
3 content solutions	6 implemented (2 decommed)

FIGURE 5.1

The content strategy team at Intuit grew from zero to five team members over two years. Adding more members helped them meet demand for the team's expertise as they multiplied partner teams, business units, and as the requests for content strategy work continued to increase—all measurable indicators of the team's success.

Schmich's content strategy team focused on problems that caused friction and those that came at a cost to the business—things like breaking down publishing silos, identifying suboptimal content management systems and processes, and decommissioning content systems, consolidating, and migrating content—all with the overall goal of improving content quality.

"Intuit couldn't wait any longer for content strategy. The cost to the business of not doing it (measured in task completion, call drivers, search retrieval and ranking, cited reasons for NPS detraction, helpfulness ratings, etc.) outweighed the cost of investing in it."

TURF WARS

When I tiptoed into the world of tech, I naively believed that turf wars like those I witnessed between editorial and advertising during my years as a journalist didn't exist among those who produced the types of digital products and services I consumed and later helped to create. But those turf wars *do* exist, often between UX content strategy practitioners and other content partners, including marketing and social media creatives, and potentially any third-party content creators your organization may contract with from time to time.

I've spent many long (and sometimes, fruitless) hours trying to align on a project with marketing content teams, only to have the efforts fail because the focus shifted from alignment to alienation—from who "owns" the content, to who is doing content strategy the "right" way, rather than focusing on how we might work together toward a common goal—to create an end product or experience that is right for the user.

Maybe the ambiguity in the early years about what content strategy actually was and where it belonged organizationally contributed to the taking of sides. Yet I find that some of my UX peers are still having to battle over things like who owns the style guide, whether or not there should be *different* style guides for different content teams, and if there should be exceptions to the style guide, depending on which team is generating the content and where and how the content is being used. I've made some strides in situations like this by focusing instead on what I call the *monster in the room*—the enormity of a project, an aggressive timeline, or dealing with a demanding stakeholder. When we put our focus there, rather than on the team we're trying to partner with, it helps us treat other content creators as partners, and to have that treatment reciprocated.

In the end, while some may try, no one team does (or should) hold complete dominion over content strategy. Given the variety of roles involved in co-creating experiences with content strategy, it's truly a shared discipline. And even if your approach to the strategic work of content is different, your users will benefit greatly if you can find a way to work together to create seamless experiences.

One of the ways your practice can extend its value and reach in a measurable way is to set up a regular cadence for office hours where anyone can bring a content question to the members of your practice. Those who sign up for a consult usually walk through an experience, wireframes, or flows to provide context, so that content strategy practitioners can noodle on solutions in real time.

Once set up, keep a record of sign-ups and note whether or not there's an increase in attendance or demand as you get more established. All of these count as quantifiable interactions and can be used to gauge practice success over time.

Educational Experiences

If your practice is on the smaller side, or you're just starting to scale practice operations, consider offering monthly or quarterly opportunities to share the discipline and the purpose of the practice with other colleagues or departments. Examples include hosting an interactive lunch-and-learn, where you share a short presentation, such as a specific use case that highlights the work of the practice and leave time for Q&A. Or you might conduct internal roadshows, booking time with different departments or business units to share what the practice is, what it does, and what it can do for the department or team you're presenting to.

Creating a standard, reusable deck in a format that allows you to add a slide or two that's specific to the department or team you're speaking with can save you time and prevent you from having to recreate the proverbial wheel every time. Generating interest is generally easier if your workplace hosts regular lunch-and-learns, but if this is new territory for your agency or organization, don't worry. Sometimes, the novelty is enough to generate interest. And if you can get a sponsor or stakeholder from your leadership team to promote the event, that's even better. Start by hosting roadshow presentations for teams you know you'll soon be working with and who are new to the discipline. Then, eventually, you can be proactive and offer to take the roadshow to other business units.

You can also do a bit of research to learn how to get on the agenda of monthly departmental meetings that invite teams from other practices or departments to give a presentation. Another way to get the word out about what the practice is up to is to create a repository for short videos that can be played on demand, where you introduce the

DEMONSTRATING PRACTICE SUCCESS AT MCAFEE

Barnali Banerji, Design and Research Senior Manager, McAfee

Demonstrating and measuring the value of content strategy has actually come easily for Barnali Banerji's team at McAfee. The first strategist to join the team developed a best-in-class approach to information architecture using the results of research on users' mental models around security offerings, along with top task analysis work. The next step was to conduct usability testing to learn if it would improve discoverability and engagement. "We then built the IA into the product for a small cohort, and the analytics for adoption, discoverability, and usage increased."

The second content strategist to join Banerji's team came onboard just as the organization was undergoing a brand refresh. Part of that effort involved conducting audits across similar screens to determine the level of consistency in content that included terms of use and privacy policies, where there was a lot of inconsistency.

Based on that discovery, the content strategist worked with the design system team to create a component for the privacy policy and similar elements, which led to an increase in the velocity of development. "We were about to capture metrics from our engineering friends and show the value. Here's the money saved, and all of the time saved,

discipline or walk through a short use case. With all of these options, you can count on increased engagement toward practice success.

Project Tracker

One of the easiest (but sometimes overlooked) ways to quantify the work of the practice is to track projects. In the earliest stages of practice-building, this may seem like overkill. Or you may think recalling the work performed in those early days will be easy enough, especially if it takes a little time for the practice to gain traction. And then all of a sudden, project requests are coming in hot, you're scrambling to bring in more resources, and you're bogged

from having content strategy create reusable components for content which can be leveraged, plugged, and played."

Content strategy also had a measurable impact on the organization's localization efforts. By predefining the amount of space needed for headers, creating reusable elements like call-to-action buttons, and determining how much text would be allowed for English in order to translate into other languages, content strategy ultimately helped to improve time to market. "The content strategists have the time, the bandwidth, the knowledge, and the know-how to actually sit down and do this. They're working with engineering. They're working with the localization teams. And they're doing the pre-work, which will then form the guardrails for the content designers to go in and do their work. It speeds up velocity of time to market. In an Agile organization, that's a big deal."

When product and content design are focused on how a particular product or feature is performing, the value and success of content strategy is measured differently from other disciplines. "Content strategists are thinking at a different level, and their impact has to be measured differently. The strategy part is more about whether you are setting things up in a way that allows content designers or product marketing to go in and use these guidelines and frameworks to do their work."

down with trying to find the right tools to help you get more work done efficiently. As a result, any "spare" time you had to track practice involvement in project work has slipped away, and with it, the accuracy of remembering exactly what was done, and how the practice helped achieve project goals.

If you make project tracking a priority in the early days of practice operations, you'll find it much easier to populate and maintain as you scale. The task becomes even easier if you've also created a practice roadmap, because then you can document projects from ideation, to implementation, to completion. Use a spreadsheet or project management software to keep track of the projects that the practice has completed.

Also, as the quantity of completed projects grows, you can extract more meaningful measures from your tracker, such as the number and variety of projects that the practice takes on or supports over a period of time, including new builds, redesigns, or platform migrations. Or you can keep a record of the number of pages, screens, or end-to-end flows impacted by the practice. You can even track the results of a content-focused bug bash, where the goal is to identify typos and other content errors that need to be fixed. Again, these measures may seem insignificant or a bit too granular (if not downright silly). Trust that when taken together and reported out, quantifiable results like these really *do* matter.

A Few Words on Failure

Failure is a tricky thing. Some of us are taught to avoid it at all costs, while others embrace failure and learn from it. Adding to the confusion are the Agile and Lean cultures within the software industry that have made it a mission to "fail fast and fail often." But what about when your practice experiences failure? What then? First, because content strategy is a cog in a bigger wheel, it's doubtful that project-level failures will ever fall completely on the shoulders of the practice. Saying this is not to point fingers elsewhere, or to shirk responsibility; instead, it's meant to remind you that the practice you've built doesn't exist in a vacuum. Is it possible to miss a product launch target because of a delay in content? Sure. Might you miss typos and other errors during QA that actually get seen by the public—and sometimes worse, leadership? Most certainly. But it can take a lot more than that to fail so completely that your practice dissolves.

So what might failure look like at the practice level? And what can you do about it? Here's the thing: it's not as complicated as you might think. At its simplest, failure at the practice level usually happens when the practice hasn't followed through on commitments made, or continually fails to meet goals. Many times, it's a lack of (or unclear) policies or procedures that are at fault. Failure can also result from setting unrealistic goals. Organizations that tend to be more autonomous are sometimes wary of establishing policies and procedures, but most organizations, whether autonomous or more traditional, can get on board with the concepts of content workflow (how the content work gets done) and governance (who makes decisions about the content), which help you get clear on content roles and responsibilities.

Short of missing one hundred percent of the goals and OKRs you've established, catastrophic failure at the practice level is unlikely. But if for some reason you feel like things are not going as well as they should, this is the perfect time to revisit the blueprint components to ensure that you haven't missed any steps. Now is the time to take stock of the resources and tools your practice has at the ready and consider whether or not it's time to reequip, reorganize, revise, or modify practice tools and resources, which are covered in more detail in Chapter 7, "Retooling." This is also a good time to look for workflow gaps and consider ways you can shore things up to keep the practice moving forward.

It may sound trite, but really, the biggest failure is in giving up on building a practice before you start, or when you meet resistance as building gets underway, for example, when you encounter tension and compression (introduced in Chapter 1 and covered in more detail in Chapter 3). If, for some unavoidable reason, the practice you've built should be disbanded or absorbed into another team or department (usually as a result of an organizational change), know that the work you've done to educate cross-functional teammates, departmental partners, and leadership will go far as they continue to contend with content.

NOTES FROM THE FIELD

RESIGNING YOURSELF TO THE STATUS QUO

Jen Schmich, Senior Manager, UX Writing, Systems & Infrastructure, Spotify

I feel like failure is just not trying—just resigning yourself to the status quo and saying, "I'm going to live with the way things are and not question it. And I'm going to accept *no* as an answer from the powers that be." To me, that's what failure really is. If you're doing something, you're going to learn something, or you're going to plant the seeds for something better to happen later.

A Little Thought Leadership Goes a Long Way

If you've been showing and sharing your work within your agency (both with internal partners and clients), or you've taken your show on the (virtual) road to different teams and departments within your organization, now's the time to consider sharing more widely. Many agencies and small-to-medium companies are open to (and sometimes encourage) employees to speak at conferences and meetups to showcase case studies and to demonstrate thought leadership as a way to foster engagement with (and increase awareness of) the company, agency, or brand. Larger organizations or enterprises might have more stringent rules around this type of participation, but it certainly doesn't hurt to ask for permission to put yourself out there, thus positioning the enterprise more prominently in the public eye.

Whatever the case where you work, if you or members of your practice have the opportunity to speak, virtually or otherwise, the external mentions you receive can have internal implications as leadership is made aware of how the practice is perceived among other practitioners, as well as among other industry players.

Speak Up and Speak Out

For those interested in speaking, but who may not feel comfortable with the idea of standing on a stage in front of a large audience, start small. Work on opportunities to get comfortable with presenting to peers first. Then work your way up (and out) to local meetups or networking groups. If there aren't any content-centric groups to speak to, there are many UX practitioners, including visual designers and researchers, who are curious about content strategy, but don't have direct access to practitioners in their agencies or organizations. A short email to propose a talk or panel discussion is a great way to share your passion for the discipline, while educating others who want to better understand what UX content strategy brings to the table. After local meetups, consider pitching a talk to a local or regional conference. Follow proposal guidelines and make sure to pitch a talk that is thematically appropriate for the conference theme and attendees. After that, if you or other members of the practice find that public speaking is something you enjoy, consider speaking at larger events. You'll find a list of content-focused conferences in the Nuts and Bolts section that follows.

If you've been speaking internally within your agency or organization, and you've shared your passion for the practice at meetups and similar networking events, now may be the time to consider submitting talk proposals for some of the bigger content strategy conferences and related events. This list is not exhaustive—not by a long shot—but it should give you a great head start.

- Button: The Content Design Conference (**www.buttonconf.com/**)
- Confab: The Content Strategy Conference (**www.confabevents.com/**)
- Design + Content: The Conference for Designers and Content Strategists (**https://content.design/**)
- LavaCon: The Conference on Content Strategy and Tech-Comm Management (**https://lavacon.org/**)
- UX Writer Conference: For Writers in the User Experience Field (**https://uxwriterconference.com/**)

The COVID-19 pandemic has opened up many virtual speaking opportunities that may have been out of reach previously, so don't be shy about casting a wide net when looking for speaking prospects.

Write It Out

Thought leadership isn't just about public speaking, which is understandably uncomfortable for some people. Instead of putting yourself on a stage or behind a digital dais, provided you have the support and permission of your company to do so, consider contributing to the greater content strategy community in written form. Again, as with public speaking, start small. If your organization has an internal blog where a post about the practice would be appropriate, put your writing chops to good use and contribute a piece to be shared across the enterprise. Also be on the lookout for external calls for white papers and similar submissions, perhaps from associations and professional organizations that are content focused. These types of pieces are likely to be shared on social media, and if nothing else, they will contribute to the aggregate number of mentions of your company or brand—a metric often counted by those who manage social media and corporate communications, and another measure of practice success.

This type of public engagement, whether through speaking or writing, isn't required for your practice to flourish, but if you and your practice peers do decide to go down that path, the internal props and accolades you'll receive can most certainly be counted toward the success of your endeavors.

NOTES FROM THE JOB SITE

#TEACHWRITESPEAK

 I love to share a good story. I'm usually more comfortable sharing in a written format, but I've found—quite accidentally—that speaking seems to be a natural progression after I've written about a topic. Consider my foray into public speaking, which was both unintentional and accidental. I attended a conference just over a decade ago and left some quite detailed (written) feedback about it on a post-conference survey. The organizers asked if they could call me to discuss my survey responses in more detail. By the time the call was over, they had invited me to speak at the conference the following year.

It was another five or so years before I spoke again, and at that time, I followed my own advice and started small at a local meetup, where I copresented a content strategy 101-themed talk with a former colleague. Around that time, my then-manager set a goal for me to speak and write as part of my performance review—a task I narrowly escaped by taking on a role elsewhere. As luck would have it, I soon revisited that task as the lead of a content strategy practice where we'd implemented lunch-and-learns and several practice roadshows, not to mention the many internal presentations I had to give as part of our efforts to gain alignment with other teams and departments.

Eventually, I started freelance writing again as a side hustle during leaner employment times. I even created a hashtag—#teachwritespeak—as an extension of my public speaking persona. As more people read my articles, the invitations to write and speak increased in frequency, especially during the COVID-19 pandemic. I've been fortunate to speak at conferences I once dreamed of attending, and often have crafted those talks from articles I've written, or about things that I'm passionate and curious about. I hadn't set out to do any of that—to gain notoriety, or likes or follows. And—real talk—I certainly hadn't sought out to write a book, but as the saying goes, the rest is history. I'm happy to be here, to have you hold this book in your hands, and to share what I've learned.

The Punch List

Content professionals who come from writing or other creative backgrounds might prefer to avoid data and statistics—really anything having to do with numbers—but to establish success measures at scale, you're going to have to nestle up to a numeral or two.

Luckily, there are many creative ways to go about creating success measures using methodologies and approaches introduced in earlier chapters, including the following:

- Journey Map
- Service Blueprint
- Process Framework

But what about failure? It's unlikely for your practice to fail to the point of extinction unless you somehow miss hitting every goal or OKR that you've established. Reworking the blueprint and using some of the suggestions for retooling can help you correct your course if the practice seems headed in the wrong direction.

And when things are going well, and you've successfully lunched and learned, or written posts for an internal blog or company intranet? You can count speaking at meetups or conferences or writing for third-party publications as measures of success for the practice as well.

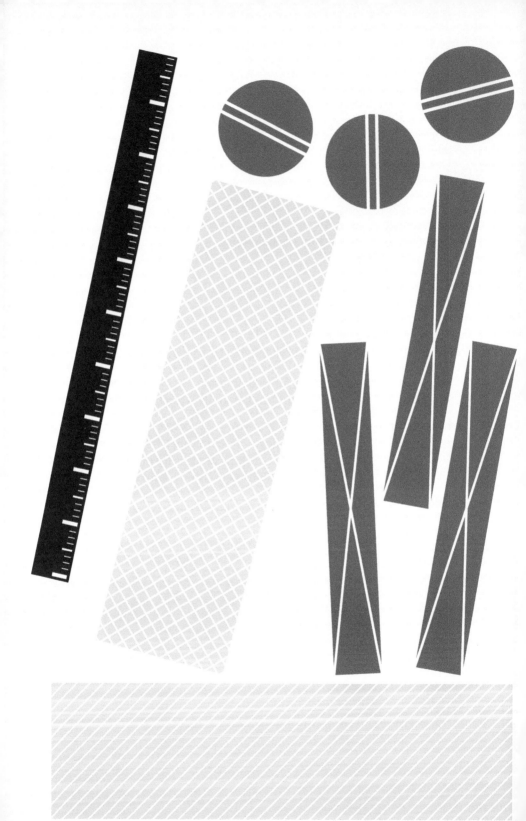

CHAPTER 6

Maintaining a Strong Core

At the center of most high-rise structures is a vitally important component known as the *core*. In the simplest terms, the purpose of the core is to help buildings withstand lateral forces, often called *horizontal loads*, that can result from wind or seismic activity. Additionally, an outer frame made of steel works in conjunction with the core to help buildings resist *vertical loads* that may result from tension or compression, which you learned about in Chapter 3, "Building Materials." Ultimately, a strong core helps to prevent collapse.

This chapter will consider ways to maintain the core of your content strategy practice, with a focus on the people, systems, and cultural environment that uphold and support it. You'll learn how to help practitioners avoid burnout, while still doing their best work. You'll also learn the importance of maintaining the systems used to facilitate the work of the practice—not just the technical systems, but the processes needed to ensure that practitioners have what they need to keep up with growing demands for their contributions, as well as advances in the discipline. Finally, you'll consider the cultural environment of the organization where your practice has been built, with a lens toward some of the resistance you may encounter when trying to stand up and maintain the practice. And you'll find out how you can safeguard the practice against that resistance.

Don't Fall Down

As it turns out, *Why Buildings Stand Up*, the volume that inspired the building and construction metaphor used throughout this book—has a sequel. Suitably titled *Why Buildings Fall Down*, coauthors Mario Salvadori and Matthys Levy explain, "The accidental death of a building is always due to the failure of its skeleton, the structure...Once you learn how structures behave, you will also learn that as if they had a social duty towards us, structures always do their very best *not to fall down*."[1] Just as the ongoing maintenance and support of a building's core is vitally important to preventing structural failure, the same holds true for the practice you've built.

The type of core maintenance spoken of in the context of building a sustainable content strategy practice includes routinely inspecting

---

1 Matthys Levy and Mario Salvadori, *Why Buildings Fall Down* (New York and London, W.W. Norton Company, 2002), 14.

the practice for any gaps in process or weak points, and it's even more so about prevention—about what it takes to keep the practice core strong *before* weakness (and the potential for failure) ever becomes a threat. Let's take a closer look at the sum of parts that comprise the anatomy of the practice you've built—the people, the systems, and the cultural environment—and consider preventative measures that can strengthen each element to help the practice remain standing, even under duress.

The Anatomy of the Practice Core

In a building, a concrete core and outer steel framework work together to help the structure withstand a variety of vertical and horizontal forces. In the human body, the central nervous system is protected by the sum of parts that comprise the human core—the brain and the spinal cord. Imagine the core of a content strategy practice as having an inner core composed of its people or practitioners; systems to help the practice (as well other business units) support the overarching goals of the organization or agency; and an outer core which is comprised of the cultural environment where the practice has been built. Figure 6.1 is a simple visual diagram that depicts how these three elements work together.

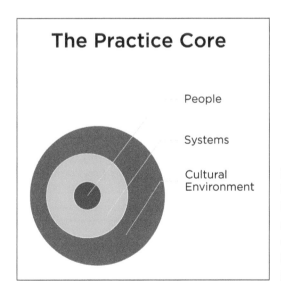

The Practice Core

People

Systems

Cultural
Environment

FIGURE 6.1
Much like the core of a building, the components of the practice core work together to help prevent the structure from collapsing.

- **People:** The foundation of a practice is built upon the collective expertise of the practitioners who do the heavy lifting of content strategy. Without the people who advocate, educate, and articulate the value of content strategy, you have nothing. As such, the people at the core of your practice must absolutely be recognized for the value they bring, and they must also be supported in a way that allows them to thrive, and not merely survive.

- **Systems:** The tools and procedures that practitioners use to do their work and achieve results are part of what comprises a system. More than just a collection of hardware and software, systems also include the business processes and operational procedures that have been put in place to help the practice operate smoothly. The systems at the core of your practice must keep pace as the discipline changes and also meet the demands of the agency or organization. Systems also need to keep pace as practitioners upskill and broaden their knowledge of content strategy. This part of the core might also influence decisions about where the practice best fits organizationally speaking.

- **Cultural Environment:** The final, outer ring that surrounds and encompasses the people and systems that comprise the practice core is the cultural environment where the practice has been built. The cultural environment is influenced by the rules and guidelines put in place by leadership: the acceptable norms and behaviors allowed and encouraged (as well as *dis*couraged) in the workspace (both physical and virtual). The problem is that sometimes those norms go unspoken until someone deviates from them, which makes it tricky to navigate the cultural environment without making some missteps.

While there is certainly some interdependency *between* these elements, if you spend time strengthening the core from the inside out—starting with the people at the center of the core and working outward—not only will you help keep the core of the practice strong, but you will also start to have more of an impact on the outer rings, from the systems to the cultural environment.

People

Investing in the practitioners who are the literal heart and soul of the practice you've built is vitally important for maintaining a strong core. Whether you are a solo practitioner, or you lead (or are part of) a team of practitioners that's been scaled to fit the demands of a larger agency or organization, the success of the practice has a direct correlation to the health and prosperity of those who are working tirelessly to champion the creation of exceptional content. Success in this context isn't just about bigger salaries or fancier titles. It's about making sure that *every* practitioner is working in an environment that is psychologically safe, as well as one that is diverse and inclusive. It's also about ensuring that each practitioner is working to their strengths and is able to clearly see how their talents and contributions add value to the practice and contribute to the success of the agency or organization.

Success also means that the practitioners aren't working to the point of burnout and beyond. Burnout is both common and insidious. So while an increase in demand for practice services can signal that there is both acceptance and understanding of the purpose of the practice (and also may be a sign that it's time to scale practice operations), if that demand results in people being spread too thin to maintain a healthy amount of work, the risk of compromising their health and safety becomes all too real.

It's possible that learning about the importance of self-care wasn't quite what you thought of when you first saw the words, "investing in practitioners." You may have expected to find recommendations on upskilling practitioners by providing training, or budgeting for professional development workshops and conference attendance, all of which are important. But the type of investment being spoken of here is more about creating space for each individual to define and explore what self-care looks like for them. Creating a safe space where each practitioner can be intentional about their well-being in a way that energizes them is vitally important to the health of the team. Here are a few suggestions for exploring how to create and hold space that encourages self-care.

Status Check-Ins

Whether practitioners are embedded in product teams, or the practice has been established as a service provider to UX, development, and other departments, having a regularly scheduled check-in for practitioners to come together to share status updates and talk craft and shop is vitally important. It's recommended that there be a shared agreement that this space is more than just a place for status check-ins—that it is also a safe space to not only share wins, but perhaps more important, it's a place where frustrations can be shared as well.

PEOPLE FIRST: THAT'S THE STATEMENT

Candi Williams, author and Head of Content Design, Bumble

Before joining Bumble, Candi Williams asked a lot of questions to ensure that the organization was a good match for her. She asked about psychological safety and well-being. And she asked about equity and inclusion. "They care about the things I care about, and that's crucial to me in any organization where I work."

In her years of experience as a content leader, Williams has learned that the people who comprise a practice or team *always* come first. "Content is at the heart of every product and experience. Yet having to explain and justify its value is exhausting for the wonderful folks who give products meaning, clarity, and relevance. Content designers and UX writers: look after *you* first. *You* matter most."

Williams has intentionally made space for her team to practice self-care by setting a standard. "I role model self-care as much as I can." This means she not only advocates for others to take time off, but she does so herself as well. Also, if she gets an inkling that people are working over their hours, she makes sure they understand that it's not something that is encouraged or expected.

Many times practitioners who are embedded on product teams, or within separate lines of business, experience friction with other teammates over things like not adhering to the agreed-upon process framework, or worse, excluding content strategy practitioners from the early phases of product development, and only involving them at the end, usually via a terse request for copy that goes something like, "All you'll need to do is add copy before we launch." There's nothing quite so frustrating to a content strategist, especially when time and effort has been invested in painstakingly building a practice. But it happens.

Some cross-functional or departmental partners can't (or refuse to) elevate content to a place of importance *throughout* the product development process. They don't understand how absolutely infuriating it can be to be brought into a project at the tail end, absent the proper context and under the pressure of having to launch on time. When that happens, practitioners need a safe space to vent and to

Another approach Williams takes to maintain a strong core is through goal setting. "I've set some collective goals about people's well-being and psychological safety and making sure that we're building an inclusive community of content designers, which is really important."

Williams works vertically and horizontally with the practitioners on her team to understand their different styles. "What are the areas you're really strong in or that you really enjoy? What are the areas that you would like support in or want to work on? Having those open and honest conversations are really important."

While the Coronavirus pandemic has forced employers and employees to have more conversations about work-life balance, as everyone was quite literally trying to establish work-life balance, there was a narrative with a primary focus on homeschooling and what to do if you had children. "I remember there were a lot of comments made to me like, 'You're lucky you don't have to worry about any of this.'" Williams considered some of that narrative to be quite one-dimensional. "What about people who are primary caregivers, or people who are living alone? There's so many different nuances. I think we have to be really mindful when we're talking about well-being, work-life balance, and self-care."

problem-solve with people who "get it." Safety in this space is key. While talking "smack" isn't encouraged, if people do need to let off a little steam, it's best to do so in a space where "what happens at status check-in stays at status check-in."

NUTS AND BOLTS SELF-CARE: DEFINE YOUR OWN BLISS

When I'm not herding content, I moonlight as a private yoga teacher. I incorporate mindfulness, breathwork, and asana in the sessions I teach in order to bring about calming and centeredness. Modalities like yoga and meditation are often included at the top of self-care lists, and while that's a good thing, it may not be *your* thing, or *the* thing for others in your practice.

In other words, what counts as self-care for some won't be the same for others. For some, caring for self might include indulgent bubble baths and similar treatments, and not so much about yoga, meditation, or similar mindfulness exercises. And self-care can also be about getting lost in something—a book, creative writing, gaming, a favorite movie, meaningful movement, restoring a classic car, even a nap—whatever it is that helps you restore yourself at a cellular level. What's most important is that each person is allowed to find—and define—their own bliss.

The Balance Score

Another way to encourage self-care among practitioners is to acknowledge the "self" in each person. Like most people, your priority for showing up to work every day is to get things done. But as the COVID-19 pandemic threw us into virtual rooms that could be easily infiltrated by a weary toddler in need of a nap (or a family member or housemate in need of pants), we've been made to acknowledge that we are so much more than the work we do, and that it's important to take time to recognize the humanity of those we work with. This is where a concept called the *Balance Score* comes in handy.

The Balance Score is a simple numeric value from 1–10 that represents how a teammate is feeling about what is on their plates in that moment, both personal and professional, and how well they feel they can balance it. Henry Poole is the cofounder of CivicActions, a partner company on a project I worked on. Writing about the Balance Score in Medium,[2] Poole explained, "A balance score is a numeric representation of how balanced we feel in the honoring of our priorities at a given moment."

Poole explains that balance in this context isn't just about being happy. He also points out that sometimes folks can be imbalanced when things appear to be OK. Instead, Poole writes, "Balance is about self-recognition of our own priorities and energy." It's also about recognizing that we are more than just the work we do and the tasks we work on. We are humans dealing with very human issues that directly impact the work we produce.

Here's how it worked in practice: During our daily stand-up calls, each team member would begin our status update by giving a numeric value that represented how they were feeling that day about

---

2 Henry Poole, "Improving Scrum Team Flow on Digital Service Projects," *CivicActions* (blog), Medium, August 15, 2019, https://medium.com/civicactions/improving-scrum-team-flow-on-digital-service-projects-6723d95eaad8

everything they had to tackle on the work front *and* on the home front. They had the option to give as much or as little detail to personal issues as they wanted. Then they would give an update about project tasks and note whether they had any blockers that would prevent the completion of those tasks. Even on the days I didn't want to go into detail, it was a relief just to be able say, "I'm handling some family stuff, and it has me a bit distracted today," and to have my colleagues not only *hear* me, but to *see* me, and to recognize the humanity we *all* brought to the table every day.

I found that there was power in not only sharing my own score out loud, but also in hearing other people's scores. We were reminded in those moments that we are people, not just practitioners from different disciplines, and even on those days when my own score was on the mid-to-lower end of the scale, it was a huge relief to know I didn't have to hide whatever was going on, and that I could bring (and be) my whole self to work. I experienced an unmatched level of empathy for (and trust in) the folks on our team, and I'm pretty convinced that I was a better colleague for it.

POWER TOOLS BALANCE SCORE BASICS

The following basic points from Poole's article can help you both introduce and model this methodology for your team:

1. The Balance Score uses a scale of 1–10, with 1 being low and 10 being high.
2. The score is composed of inputs from three areas, or contexts, as Poole calls them: personal, professional, and spiritual.
3. Poole writes that balance scores can communicate two numbers about priorities: how well you know what they are, and how well you are able to honor them.
4. Balance Scores can also do the following:
   a. Signal a shift in priorities.
   b. Cultivate connections with teammates.
   c. Help eliminate finger-pointing and making assumptions when things go wrong.
   d. Shift the focus on the *people* doing the work to help improve performance.
   e. Help to foster trust and relationship-building among teams.

While the Balance Score has origins in Agile ceremonies, it can be easily adapted into any gathering of practitioners where status updates are given.

Vulnerability as a Superpower

It's impossible to have a conversation about vulnerability without mentioning the groundbreaking work of author and researcher Dr. Brené Brown, whose TED talk, "The Power of Vulnerability,"[3] consistently ranks among the most viewed and most popular among the conference organization's "ideas worth spreading." There's no single or simple way to break down Brown's years of work on this topic, although her 20-minute TED talk does a more than adequate job of introducing and distilling it into three areas of commonality she found in her research among people whom she classified as "wholeheartedly vulnerable":

- **Courage:** Telling your story with your whole heart
- **Compassion:** Being kind to yourself first and then to others
- **Connection:** Accepting who you are over who you think you should be in order to connect authentically with others

In a message to corporations Brown said, "We just need you to be authentic and real." And therein lies the crux of the matter—how the willingness to be vulnerable in the workplace can have a domino effect on others. And that willingness starts with self-care and self-acceptance.

You should understand that vulnerability—and, in fact, any of the modalities discussed here—are not meant to be prescriptive or enforced as mandatory. Instead, they are meant to provide pathways toward helping each person in the practice define their own expression of self-care, and to know it's not only permissible to figure out what that means to them, but that it's encouraged and supported, too.

Systems: The Content Strategy Process Overview

If self-care focuses on how practitioners work together with each other and within teams, systems focus on how things get done. By way of setting expectations, this is not meant to be a "how to design content systems" discussion. There are many books that cover topics like how to choose and set up content management systems (CMS)

---

3 Brené Brown, "The Power of Vulnerability," TEDxHouston, June 2010, www.ted.com/talks/brene_brown_the_power_of_vulnerability?language=en

and related technologies. Nor is it meant to be a guide on how to use collaborative design tools like Adobe XD, Figma, or Sketch—all quite popular at the time of this writing, and all with ample documentation available online, along with many hours of YouTube tutorials that dive deeper into how to use these tools. As introduced in Chapter 3, there are many authors whose work involves systematic approaches to the discipline of content strategy, including authors Halvorson and Sheffield.

Instead, this section introduces the content strategy process overview as a system to help set expectations for both internal and external clients and stakeholders, and to share what they can expect to encounter as the work of the practice unfolds: the components of a content strategy kickoff, the phases of content strategy, and a sample list of deliverables that might be included as the project unfolds. While this overview gets a bit into the "how-to" territory, its purpose here is to provide a few different systematic approaches to help maintain the core of your practice.

The Content Strategy Kickoff

The purpose of the content strategy kickoff is to give practitioners dedicated time with clients, product owners, or stakeholders to better understand the goals of the project and to extract content requirements that will inform the overarching content strategy. Ideally, this process will happen shortly after the overall project kickoff so that content strategy can get in lockstep with the overall product development process. The content strategy kickoff includes:

- **Business Objectives**
  - Define business goals.
  - Define digital strategy goals (including web, apps, and social media, as applicable).
  - Establish message architecture and information hierarchies that convey what information or ideas will be provided for users or customers.
- **Heuristic Analysis**
  - Review current digital experience(s).
  - Check heuristic results against analytics to identify corroboration with known or suspected pain points.

- **Customer Research**
  - Understand user goals.
  - Develop personas.
  - Get into user mindset.
- **Content Analysis—Deep Dive**
  - Define content ecosystem: What content channels exist?
  - Audit content ecosystem: Include a full or representative sample, including social media and other touchpoints as applicable.
  - Comparative analysis: Review content within the digital experiences of true competitors or industry peers.
  - Mind the gap: What content is missing that your users might need or want?
  - Readiness: How prepared is the client or organization to receive, implement, and maintain content strategy recommendations?
- **Analytics**
  - Deeper dive into analytics.
  - Identify relevant content KPIs.
  - Understand user interactions.

The Content Strategy Phases

Content strategy phases may include some or all of the following. (Note: the names of each phase will vary depending on project scope, type, and relevancy):

- **Proposal:** Contribute to the creation of project proposal if needed. Create messaging architecture (with stakeholders) to discover content opportunities if time allows.
- **Discovery/Analysis:** Review Statement of Work and interview notes to understand stakeholder objectives, user needs, and proposed (design) solutions. Conduct content audit to determine content needs.
- **Design:** Focus on what existing content can or will be repurposed, what needs to be created from scratch, who will write the content, editorial standards, and processes?
- **Build:** Create new content, repurpose existing content, and collect/edit content curated by clients.

- **Test:** Give QA/Test team guidance on what content work was done (what's new, what changed, functionality, etc.), and create a list of specific content changes to verify that all content work made it to the finished product. Also develop a plan for reporting content defects.

- **Maintenance:** Develop a plan for creating new/updated content, create editorial calendar (if applicable) based on business needs, and removal or archival of outdated content.

The Content Strategy Deliverables

This collection of content strategy deliverables has been curated from a variety of sources[4] and is provided here as an example. The deliverables produced by your practice will vary depending on project scope, type, and relevancy, and could include some or all of the following:

- **Discovery Phase Documents/Deliverables**

  - Content Project Summary
  - Content Inventory
  - Content Audit

- **Analysis Phase Documents/Deliverables**

  - Content Strategy Findings and Approach
  - Existing Content Analysis
  - Gap Analysis
  - Comparative Analysis
  - Editorial Process Analysis and Recommendations
  - Resource Readiness (People and Systems)

- **Design Phase Documents/Deliverables**

  - The Content Matrix
  - Editorial Style Guide
  - Content Approval Process
  - Content Translation Process
  - Content Management System (CMS) Consultation

---

4   Kristina Halvorson and Melissa Rach, *Content Strategy for the Web*, 2nd ed.; Kevin P. Nichols, *Enterprise Content Strategy*; Richard Sheffield, *The Web Content Strategist's Bible*.

- **Build Phase Documents/Deliverables**
  - Project Tracker
  - Content Review and Approval
- **Maintenance Phase Documents/Deliverables**
  - Content Maintenance Process
  - Editorial Calendar
  - Content Removal Process
  - Content Archival Plan and Policy

The Content Lifecycle

It would be nothing short of blasphemous to talk of systems in the context of content strategy without mentioning the content lifecycle. It was first introduced by content strategist, Erin Scime, in or around 2009, and later included in *The Content Strategy Toolkit* by Meghan Casey. The content lifecycle includes the following five phases depicted in Figure 6.2:

- **Assess:** Assess content performance against current or future-state goals, depending on the goals of your project.
- **Strategize:** Resulting from the heavy lifting of content strategy discovery, this phase is where you identify what content is needed, by which users, and how it will be presented.

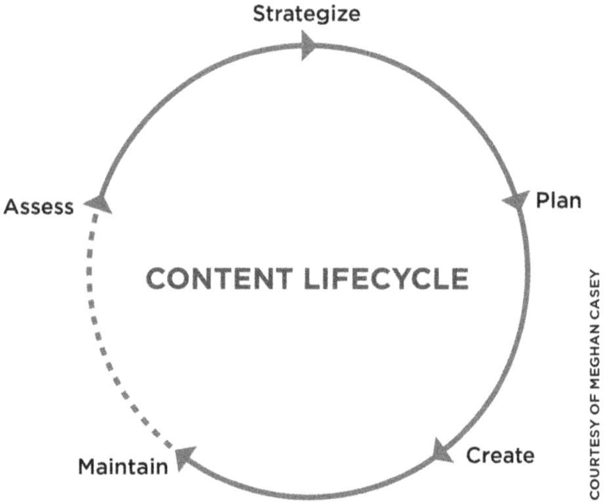

FIGURE 6.2
The five phases of the Content Lifecycle help to establish a systematic approach to content that supports changes in business goals or user needs.

- **Plan:** This phase includes defining content roles and responsibilities, as well as the processes needed to support them.

- **Create:** This phase includes the people and tools needed—collectively, resources—to create, approve, and publish content.

- **Maintain:** This phase includes periodically reviewing, updating, deleting, or archiving content as business goals and user needs change.

As Casey writes, "Reassessing your strategy regularly is important, especially as your business model or priorities change, new competitors come on the scene, or your target audience shifts."[5] This lifecycle approach to content—and by extension, content strategy as a practice—is a great way to check the core strength of your practice, and to shore up any weaknesses in the systems that have been put in place to ensure the practice operates smoothly.

DesignOps and the Content Strategy Practice

Your practice may be situated as a service to other teams that impacts both user experience and marketing driven initiatives. Or it may be part of a larger UX design team. In either instance, if there is a larger group of disciplines that have assembled into a formal design operations practice—aka DesignOps—you may need to consider whether the longevity of your content strategy practice is best served by maintaining autonomy and providing *a service to* DesignOps, or by assimilating and becoming *part of* DesignOps.

DesignOps Defined

Many sources credit Dave Malouf, a leader and educator in digital design, with coining the term *DesignOps*. Even though the definition of the term has evolved over time, according to Malouf, the most basic definition is that design operations is the supporting structure—the scaffolding, if you will—of design. The purpose and mission of DesignOps, writes Malouf, is "amplifying the value of design practice."[6] Speaking on the subject of the evolution of

---

5   Meghan Casey, *The Content Strategy Toolkit* (San Francisco: New Riders, 2015), 203.

6   Dave Malouf, "Amplify," *Amplify Design* (blog), Medium, October 29, 2019,
    https://medium.com/amplify-design/amplify-84ae18edd25a

DesignOps definitions in a Rosenfeld Media podcast,[7] Malouf explained, "If design is made up of the processes, methods, and craft that we do as designers...operations is pretty much everything that makes sure that those processes, methods, and crafts are successful and valuable to both [the] user...and the organization that's sponsoring the activities." This definition parallels much of what has been shared in this chapter: the importance of communicating the value of the content strategy practice.

Another definition of DesignOps comes from the Nielsen Norman Group in an article titled "DesignOps 101:"[8] "DesignOps refers to the orchestration and optimization of people, processes, and craft in order to amplify design's value and impact at scale." In a companion piece, NNG cited the importance of framing DesignOps "as a holistic, flexible practice that can shift to meet the needs of your team over time."[9] Again, there is a parallel between this framing and building of a content strategy practice in a sustainable and scalable way. The goal of this very brief discussion on the definition of design operations is not to debate which definition is the best or most accurate. In fact, there are many other definitions out in the wild that are quite distinct from the two shared here. Instead, for those people who are building a content strategy practice in an organization that also includes a design operations practice, the goal is to consider the best organizational fit for content strategy.

Where the Practice Fits, It Sits

As the definition, mission, and purpose of design operations has matured over time, it has also become more inclusive in terms of the many specialized, cross-functional disciplines that comprise a DesignOps practice. And more often, content strategy has been explicitly called out among those specialized disciplines. As you are working on building a practice within your agency or organization, and particularly when you are focused on the second component of

---

7   Dave Malouf, "Exploring the DesignOps Definition: A Chat with Dave Malouf," *The Rosenfeld Review* 🎙 (podcast), 2018, https://soundcloud.com/rosenfeld-media/malouf-podcast?utm_medium=socialmedia&utm_source=eux_twitter&utm_content=DaveMaloufPodcast

8   Kate Kaplan, "DesignOps 101," Nielsen Norman Group, July 21,2019, www.nngroup.com/articles/design-operations-101/

9   Kate Kaplan, "DesignOps: What's the Point? How Practitioners Define DesignOps Value," Nielsen Norman Group, May 24, 2020, www.nngroup.com/articles/design-ops-definitions/

the blueprint—building strong relationships with cross-functional teams—you'll want to ensure that every member of the design operations team understands how content strategy will impact their work. They also should understand the value that a strategic approach to content brings to design operations and to the organization as a whole.

Imagining a Future, Together

Let's return to the Nielsen Norman Group's article "DesignOps 101" where the definition of DesignOps was shared and consider NNG's three goals of DesignOps:

- How we work together
- How we get our work done
- How we create impact

Whether or not the content strategy practice you're building gets incorporated into a larger design operations practice, or it maintains autonomy as a service provider to other practices and departments, you'll want to partner with those who lead DesignOps in your agency or organization to imagine your future together. This short list of goals is a great conversation starter, and it uses a language familiar to design operations leaders. If you can use these goals (or something similarly worded) to help broker conversations with DesignOps leaders, you can then build on those conversations using the techniques outlined in Chapter 2, "Structural Alignment," on building alliances to identify shared methodologies and goals. Ultimately, you can work together to figure out a model or framework for how DesignOps will engage with the content strategy practice toward the achievement of these shared goals.

Cultural Environment

In the book *Liminal Thinking*, author Dave Gray explained, "You can cultivate a way of thinking and being that will allow you to have...breakthrough insights more often," and that in doing so, "...you will be able to guide others to similar mind shifts that will give them the power to transform their lives...to create new doorways to possibilities." The book also included a quote from Marshall McLuhan that read, "Once you see the boundaries of your environment, they are no longer the boundaries of your environment."[10]

---

10 Dave Gray, *Liminal Thinking* (New York: Two Waves Books, 2016), p. xviii.

In the spirit of this quote, remember that creating boundaries is as much about defining what happens within a given space, and not so much about keeping things out. When you acknowledge the boundaries of the agency or organization that your practice is a part of, it helps you understand their potential impact upon your practice. It also informs how you can organize practice operations within operational boundaries to maintain core practice strength, hopefully with minimal disruption to day-to-day operations. Although following operational standards and procedures may not be optional, how your practice functions *within* those operational boundaries can help you create a practice space that is inclusive, safe, welcoming, and diverse—for your practitioners, as well as for your cross-functional teammates and clients, both internal and external. As you continue to do the work to nurture the practice and maintain its core strength while achieving goals and desired results, you may find that your ability to impact the outer layers of the practice core increase over time.

Build Inclusive, Safe Spaces

Your practice should ideally be reflective of the people you want to benefit from the work you produce—your users. By extension, the practice space, whether figurative or literal, should embrace the intention of inclusive design: to include everyone, regardless of differences in ability, gender, race, sexual orientation, or other characteristics, in the activities of the practice, so that they feel safe to exist as they are in the workplace without fear of discrimination. Your practice should be built to operate in ways that honor inclusivity, and to the extent that you are able, in a way to accommodate practitioners of all types.

NUTS AND BOLTS THE NOMENCLATURE OF INCLUSION

If you think that an organized group of content strategists can't impact the cultural environment of a large organization, think again. Consider the work of Intuit's Content Systems Team working in partnership with members of the organization's Racial Equity Advancement Leadership (REAL) team, who collaborated to document approaches to eliminating racist and exclusionary language from the company.[11] Their work

---

11 Intuit Blog Team, "6 Ways to Abolish Racist Language," *Intuit Blog*, February 25, 2021, www.intuit.com/blog/social-responsibility/6-ways-to-abolish-racist-language/?s=inclusive+style+guide

supports creating product and customer experiences that are more inclusive. You can see the full content design system at https://contentdesign.intuit.com/.

Be Deliberately Diverse

It's hard to discuss inclusion without mentioning diversity, which is related but not the same. In their online guide called "How to Begin Designing for Diversity,"[12] authors Boyuan Gao and Jahan Mantin explain diversity and inclusion as follows:

> **Diversity is** *quantitative.* It's the composition of different people represented in what you make, and the decision-makers on your team.

> **Inclusion** speaks to the *quality of the experience you've designed* for these diverse folks, so they experience themselves as leaders and decision makers.

Gao and Mantin also pose this critical question, to be asked by design teams that want to embrace design diversity: "How do the identities within your team influence and impact your design decisions?" They go on to write, "When we join forces with a collaborative group, we each enter with our own values, priorities, and goals—in addition to the cultural and racial identities we bring to the table."

Building diversity into your practice gives you an advantage in the achievement of diverse design goals. It demands that you expand your thinking beyond the folks seated at the product table, which, let's face it, aren't always the most diverse. Instead, it invites you to avoid doing the same things, in the same ways, with the same (types of) people. It asks you to think about who is *not* represented in your products and services, and what it will take to include them—a process that often begins with how you "speak" to users with content and how tone shifts to be inclusive for all. You've read about the impact inclusive content can have on an organization like Intuit, and you can believe that a determined group of content strategy practitioners have what it takes to get the attention of leadership around the importance of being intentionally inclusive and diverse.

---

12 Boyuan Gao and Jahan Mantin, "How to Begin Designing for Diversity," The Creative Independent, September 2019, https://thecreativeindependent.com/guides/how-to-begin-designing-for-diversity

Beware the Culture Trap

What does culture mean in your agency or organization? What does it mean in the context of the practice and the people who comprise it? In an online article titled "The Culture Trap,"[13] author Dr. KellyAnn Fitzpatrick included this quote from the book *DevOps for Dummies*: "Company culture is best described as the unspoken expectations, behavior, and values of an organization. Culture is what tells your employees whether company leadership is open to new ideas. It's what informs an employee's decision as to whether to come forward with a problem or to sweep it under the rug."

It's the unspoken part that's deserving of a warning, and that demands that, for the safety and sustainability of your practice, you take the time to learn as much as you can about the culture of the organization where you are building your practice. As Dr. Fitzpatrick wrote, the concept of a company culture can also be code speak for "gatekeeping and exclusion"—in short, the opposite of diversity and inclusion. Resistance to the creation of the practice—and, in fact, to any change in organizational structure—can be quite telling, and what you need to know in terms of navigating and understanding culture will be revealed to you as you go. To the extent that you are willing, able, and comfortable doing so, pushing the boundaries of corporate culture outward from the practice core requires strength and endurance, and in many cases, can yield impactful results at the organizational level.

This outer layer of the core—the cultural environment—is sometimes impenetrable from the outside. But if steps are put in place to ensure that the elements at the center of the core are reinforced, the practice can not only bring about a culture of content strategy competency, but it can also begin a shift to a more diverse and inclusive environment.

---

13 Dr. KellyAnn Fitzpatrick, "The Culture Trap," RedMonk, October 30, 2019, https://redmonk.com/kfitzpatrick/2019/10/30/the-culture-trap/

The Punch List

Maintaining a strong core is vital to ensure that the future for your practice is sustainable and scalable. As you're building your content strategy practice to include a strong core, you'll want to do the following:

- Make plans to avoid burnout *before* it happens by advocating for self-care, as well as continual training, cross-functional pairing, and enough content strategy resources to handle the demand for your services.

- Decide how the content lifecycle fits into the larger product development process so your cross-functional peers and partners understand that the successful completion of a client project or an enterprise's digital experience is a *repeatable* process that everyone contributes to, learns from, and gets better at over time.

- If DesignOps exists in your agency or organization, consider whether assimilating into or remaining autonomous from that practice is best for ensuring the longevity of the content strategy practice. (And yes—you can have a practice within a practice.)

- Learn what you can about the cultural environment where you're building your practice and beware of the culture trap, where gatekeeping and exclusion eclipse diversity and inclusion. Establishing operational boundaries that protect the work of the practice can maintain core strength.

One last thing: Maintaining a strong core should not require the complete teardown of a structure unless conditions are found that increase the potential for a full collapse. However, if failure is looming, you can revisit the blueprint components to make sure that you haven't missed anything critical or useful in the building process. Yet, there are still times where the dismantling of a practice does result in the reduction of practitioners, the absorption of practitioners into a different department or function, or even the complete elimination of the practice. Structural failure in this instance is usually unavoidable.

There are also times when a practice that's been completely torn down will need to be rebuilt again. If and when that happens, glean what you can about what caused the original practice structure to fail—whether outward forces or some internal structure failure—and as you work through the blueprint from the beginning, do what you can to include structural support from the outside in.

CHAPTER 7

Retooling

In Chapter 3, "Building Materials," you learned about creating a process framework to capture all the phases of site or product development that the practice is likely to be involved in. You also learned how mapping out those processes ensures that you've included all cross-functional disciplines and departments that the practice interacts with along the way in order to build alignment. Additionally, creating the process framework identifies the handoffs that need to happen *between* disciplines to ensure that the practice produces the right deliverables and reinforces the importance of engaging the practice consistently, from kickoff to launch.

Chapter 3 also included a starter set of tools for you to evaluate and adapt to projects of varying sizes. If all has gone well and you've had success using those tools, it's likely that the first few projects you've completed have brought about an increased demand for content strategy, resulting in an increased volume of work. But what if things have not gone quite as smoothly as you'd hoped? First, don't be discouraged. Second, take a moment to regroup and identify friction points, where there may be discord or disagreement with or between cross-functional teammates or departmental partners, or fail points, where perhaps you've discovered holes in the process framework that need to be addressed.

Whichever scenario resonates with the trajectory your practice is on, this point in the practice-building journey is a great time to reconsider the tools used by your practice—both existing tools and tools yet to be discovered—that will help your practice deliver content strategies at scale. This process, called *retooling*, examines whether your existing toolset needs to be adjusted, updated, or replaced to facilitate scaling and growth, both in the number of projects your practice can support and as your practice matures.

The brief introduction to retooling in Chapter 4, "Expansion: Building Up or Building Out," introduced retooling as a way to increase practice capabilities incrementally, which is useful in situations where you're unable to add more people resources to your practice. This chapter takes a closer look at the retooling process and provides you with more detail on what it means to retool, as well as when and how to do it.

Retooling is important to content strategy practices of all sizes. In addition to the starter set of tools that were introduced in Chapter 3, solo to small practices might find tools that automate processes previously done manually, such as crawlers to help speed up website inventories so your practice can handle larger projects. Mid-sized to larger practices might find tools that fill a capability gap, such as gaining access to analytics, so the practice can include making data-driven recommendations as part of an ongoing content optimization strategy. In short, retooling involves refining existing tools and methodologies to enhance efficiency and improve capabilities so the practice can quickly adapt and adjust as your agency or organization evolves in response to changes in the marketplace.

Retooling: What It Is and How to Do It

According to the Merriam-Webster's online dictionary, the definition of *retool* is, "to reequip with tools; reorganize; revise [or] modify," and, "to make especially minor changes or improvements." Other definitions point to helping people change in order to adapt as organizations change. Each component of the definition has a slightly different application in the context of building and maintaining a content strategy practice, and each one is equally important to understand:

- **Reequip:** In the context of building a practice that's both scalable and sustainable, reequipping requires taking a closer look at your current toolset to determine whether or not you need to augment what you have to meet new demands.

- **Reorganize:** Instead of adding to your current set of tools, reorganizing involves tidying up your existing toolkit to ensure that you have all your implements ready in anticipation of incoming work and to help you meet project milestones with efficiency and on time.

- **Revise:** You might say that revising your practice toolkit combines a little bit of reequipping with a little reorganization, with the goal of updating the tools you have on hand in order to improve practice capabilities, and perhaps discovering alternate uses for those tools.

- **Modify:** This approach involves making basic changes to your toolset in order to reorient your practice to take on a wider variety of projects or to add to the overall efficiency of your practice.

Whatever approach you take, remember that retooling is integral to successful growth and scaling. It is also crucial for establishing and maintaining stability, as well as making sure that your practice can meet the ever-changing evolution of digital experiences.

When to Retool

No matter which retooling approach you apply to your practice, as with the discipline of content strategy, retooling is not a one-and-done process. In fact, retooling should become a regular part of reassessing practice performance, with an eye toward adjusting practice operations to stay on pace with changes that signal a directional pivot for your clients or your organization, and to anticipate new or modified business goals that are necessary to stay competitive.

So when exactly should you retool? It depends. The cadence you settle on will be contingent upon many variables, such as the frequency of change in your industry, or as demands for new features, products, or services increase. Retooling can also be timed to coincide with any internal checkups that your enterprise might conduct, such as quarterly or annual organizational reviews. At a minimum, you should consider retooling at these critical points of change in your practice:

- **When considering growing or scaling:** You can consider it a good thing if your practice is beginning to experience growing pains, and you find that you're struggling to keep up with demands as the frequency of requests increases. Now is a good time to retool. In fact, if you have an idea of the direction that you want to take the practice by adding extra value, you can proactively retool in anticipation of those changes. For example, you might offer to create customized style guides that help clients maintain core content strategies, or provide content model documentation to in-house developers to encourage and facilitate content reuse.

- **When adding new practitioners:** Sometimes adding new practitioners increases *existing* capabilities, while at other times they add *new* capabilities to your practice. Either way, retooling as the number of practitioners grows not only calls for you to consider what new tools might be added to your toolbox, but also requires you to consider how to reorient existing practitioners to work within their strengths, and in ways that complement the skills of newcomers to the practice.

- **When practitioners leave:** If a practitioner's departure leaves a void in practice capabilities, retooling will help you figure out how to fill that void. This is especially important if you're unable to backfill a vacant role with another practitioner, due to budgetary or other constraints.

As you can see, retooling is not just for tools and frameworks—it's also for people. As client or business goals evolve—as new products are developed, or new services are offered—the practice and its practitioners will need to evolve as well. Keeping up with changes in the discipline is vital to the health and stability of your practice, whether or not it's through training, earning certifications, or some other method of refreshing skills. Remember that improving the capabilities of every practitioner contributes to the longevity and sustainability of the practice.

NUTS AND BOLTS RETOOLING VS. FINE-TUNING

Up to this point, it's likely that much of the work your practice has focused on has included projects of a limited scope, maybe stemming from an app or site redesign, or perhaps in the form of migration work in support of replatforming to a new content management system (CMS) or other technology. The tools presented in Chapter 3 are all appropriate for these types of projects, and with a little fine-tuning, they can also accommodate the delivery of content strategy work at scale.

The Practice Roadmap: Know Before You Grow

It helps to know where the practice itself is headed, both in the context of business goals, as well as by identifying content opportunities that are independent of business objectives.

Creating a practice-specific roadmap gives you an opportunity to document project-agnostic ways that the practice can ultimately support the business. For example, you can include creating a comprehensive content matrix on your practice roadmap. This matrix is where you'll document content owners, legal or regulatory reviewers, and content review dates. Tracking this information saves time and money when content comes up for review by eliminating the time-consuming (and expensive) task of hunting down this information at the last minute.

In fact, aside from knowing what projects are coming up, you want to be in control of deciding what the practice focus will be outside of product support and how best to direct practice resources. When you take a focused approach, it prevents the practice from becoming siloed, meaning that it is solely engaged in routine content strategy work. Routine work isn't a bad thing, especially if your client or organization has a large volume of content that needs attention. However, without having a focus beyond routine work, the perception of the value that the practice brings to the table can become limited. As well, establishing a practice roadmap can prevent the practice from becoming reactive, springing into action only when called upon, such as when content needs to be created or updated. Of course, you want to be in service of your cross-functional team, and you want to lend support to departmental partners as well. By the same token, you also want to position the practice in a way that allows innovation as you work proactively toward adding value in addition to or outside of project-based work.

Product roadmaps are usually created and maintained by teams that practice Agile methodology. If your organization practices waterfall methodology, you'll likely find Gantt charts or a similar tool to track projects and priorities tied to business goals over time. What this looks like in an agency will differ a bit from what it looks like from an in-house point of view. In fact, if yours is an agency-based practice, you might be wondering, how do I establish a roadmap that's not tied to a client project? Think outside the box. Create a practice-level roadmap and include milestones that capture how you want to level up the practice when you're not knee-deep in client or project deliverables. Figure 7.1 shows an example of what your practice roadmap might look like.

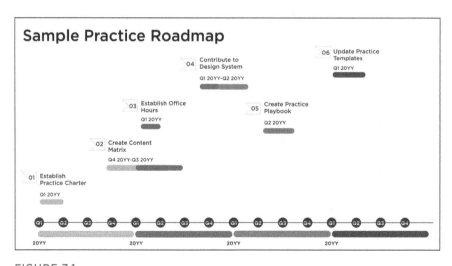

FIGURE 7.1
Agency-based practices can adopt this sample format for clients as a way to help map activities to maintain the core strategy or as a value-added service.

A content strategy practice roadmap enables you to prioritize and track content initiatives that may:

- **Create and populate a content matrix**, based on information gleaned from content inventories, and then building it out to track additional information to facilitate content management.
- **Create and modify deliverable templates for reuse with multiple clients**, or on multiple projects, with a goal of creating a central repository of practice resources.
- **Find and try out new tools and methodologies** that improve or enhance practice capabilities.
- **Build a content strategy knowledge base** for use by practitioners and others, with the goal of cultivating a culture of continuous learning.
- **Attend workshops, webinars, and meetups** to enhance and expand practice knowledge and capabilities.
- **Schedule lunch-and-learns and internal workshops** for your agency peers and partners to enhance their understanding of the purpose and capabilities of the practice, as well as how you work with each team or department within the agency (or how you can work with them in the future).

You can create a simple practice roadmap, which is basically another way to approach retooling. Even though these activities aren't tied to a specific client or linked to project-specific milestones, they can result in added value for future clients or projects. In addition, as the collective knowledge of practitioners gets broader and deeper, you'll be able to offer added value to clients who initially sign up for the more fundamental services offered by the practice. For now, you should capture this information in a running list format.

NUTS AND BOLTS CROSS-FUNCTIONAL RETOOLING CHECK-IN

When you retool, be sure to survey cross-functional teammates and other partners who, through repeated exposure to the inner workings of the practice, might have some ideas that you'll want to consider adding to your list. Project retrospectives and other cross-functional check-ins conducted with your team may provide inputs for that list as well. You can also revisit the process framework to assess how well it's working from the point of view of the team and add appropriate suggestions to your practice roadmap as they come up.

After you've listed activities for the practice to pursue, you can then plug in what you know about current or ongoing clients or in-house projects, along with any future projects that are on the horizon, into a product roadmap, a Gantt chart, or some other timeline. Then you'll slot in the items from your practice roadmap based on the list of non-project activities you created.

On the surface, this task might appear to be easy for those who have built an in-house practice. After all, your product team or organization likely has an established roadmap or timeline that you can easily access, and it helps you tie in practice priorities to overall business objectives. But again, for the same reasons that your agency counterparts were invited to do so, think outside of the proverbial box. Try to think beyond what is possible at the project level and consider how the practice can add value at the organizational level.

NUTS AND BOLTS END-TO-END, THEN DO IT AGAIN

The expression "end-to-end" seems to come up more frequently in content strategy job descriptions as a desired skill, but it seems like a bit of a misnomer. End-to-end tends to suggest that content strategy—and in fact, product development—is linear, when in fact, it's cyclical. For example, Agile methodology refers to a product development lifecycle, which includes phases that are very similar to the content lifecycle discussed in the previous chapter. The main point to remember is that just like retooling, content strategy isn't solely linear. Because content is as much of an asset as the products and services that a business creates or offers, adhering to the lifecycle enables you to manage content more efficiently across an enterprise.

The Practice Playbook: Documenting Processes for Repeat Success

As your practice grows and matures—and as the approaches, deliverables, processes, and procedures increase with that growth—storing information that's critical to practice operations in several different repositories will become unmanageable at best and completely unsustainable at worst. Creating a practice playbook is one of the best ways to track this information. As a living document, it provides a single source of truth for practice operations for you (as the practice lead or manager), your cross-functional teammates, and for your organization, that you can modify as the practice changes.

How you assemble your playbook is entirely up to you and your team of practitioners, although you may want to employ a tactic similar to how you documented items for your practice roadmap—brainstorming with practitioners, cross-functional teammates, and any other partners the practice interacts with on a regular basis. As for how you format the playbook and where the information is stored, there are many online tools you can use that will allow you to update the contents whenever necessary and give you access to multiple users and viewers. Check with other teams you work with to determine if there are other playbooks in your agency or organization that you can use as a guide when creating yours.

What you include in your playbook is also up to you. As you begin assembling the contents of your playbook, consider the audience you're creating it for, or who you may want to review it in the future, such as internal stakeholders. Agency-based practices should also consider whether sharing the playbook (or an overview of its contents) with clients is appropriate to demonstrate practice maturity and to answer any questions that potential clients might have about the capability of your practice to provide value to their project.

Table 7.1 includes a sample list of components that you can consider for inclusion in your playbook, along with a brief description of the purpose of each one. They are listed alphabetically here for easy scanning, but you can decide the order of information that's most appropriate for your situation. Depending on whether your practice is agency-based or in-house, some components may or may not be applicable to the playbook you create.

TABLE 7.1 SAMPLE COMPONENTS OF A PRACTICE PLAYBOOK

Playbook Component	Purpose
Content Management System (CMS)	High-level overview of the CMS used to publish content, along with a summary of interactions (if any) that the practice has with it. Include link(s) to more detailed documentation if needed.
Content Matrix	Summary description of the purpose of the matrix, how often it's updated, and a list of information categories it contains.

continues

TABLE 7.1 CONTINUED

Playbook Component	Purpose
Content Models	If your practice creates or uses content models to document elements within content types for web or mobile experiences, list each model by name, followed by a short description of what they're for, how they're used, and provide a link to whatever repository they are kept in.
Content Goals and Principles	Goals for content created for a client or in-house, usually in support of broader business objectives, along with a list of directives that address how the practice approaches content to achieve those goals.
Content Strategy Templates	List of templates used by the practice for various content strategy tasks and deliverables.
Cross-Functional Points of Contact	List of titles of cross-functional teammates responsible for engaging with the practice.
Deliverables List	List of all potential deliverables that could be created for a client or in-house project. Be sure to note that not all deliverables will be applicable to all projects.
Practice Charter	The complete text of your practice charter. Consider having this document at or near the beginning of the playbook, because it sets the tone for everything else contained within.
Practice Capabilities	Another list, this time documenting at a high level all of the services offered by the practice. For example, if your practice is also responsible for copywriting (implementing the strategy), or content administration (inputting copy and other content elements into the CMS) include these on the list.
Process Documentation	A description of any processes that support services provided by the practice at critical milestones defined by your practice framework, including dependencies and handoffs to other cross-functional teammates and partners.
Process Framework	Visual documentation of process framework as described in Chapter 3. It can be represented by a screenshot with a link to the actual artifact.
Project Summaries	A brief overview of projects handled by the practice. Include a brief problem statement, inputs that informed the approaches taken to solve the problem (i.e., analytics and research), and a summary of the outcome.

continues

TABLE 7.1 CONTINUED

Playbook Component	Purpose
Success Metrics (Practice-specific)	A list of metrics used to gauge *practice* success. See Chapter 5, "Measuring Success," for more information on measuring success at the *practice* level.
Tools List	List of tools, such as those outlined in Chapter 3, that may be used at the project level to develop content strategies. Use screenshots of diagrams (if available) to demonstrate visually how tools are applied.

NUTS AND BOLTS EXPLORING PRACTICE PARTNERSHIPS THAT ENHANCE OFFERINGS

Chapter 2 included a list of cross-functional teammates that you'll likely engage through the establishment of the practice, and it also cited examples of what interactions with those teammates might look like. As you implement retooling, consider whether or not there are additional partnerships that you may not have thought of in the past, either because new competencies or disciplines have been added, or because you've identified new opportunities with current teammates that can further enhance practice offerings, such as the following:

- **Visual Design:** Co-create design and content principles.
- **Instructional Design:** Collaborate on customer experience, knowledge base, or help desk content strategies.
- **Engineering:** If content modeling is part of your practice capabilities, co-work with content curious engineers on creating a repository of content model documentation.

INSTRUCTIONAL DESIGN AND CONTENT STRATEGY

Berni Xiong, author, coach, content strategist and instructional designer

About a decade before getting into instructional design, Berni Xiong was a personal coach in the professional development space, focused on creating curriculum and delivering trainings and workshops in both the public and private sectors.

Xiong wore all the hats—including content creation—to execute the service she provided. Her skills in coaching and training helped her transition easily into instructional design, but she soon discovered that her knowledge of content strategy was limited. "I was creating training content willy-nilly and then realized I'd been doing it all wrong—that there needed to be a rhyme and reason behind everything."

As her expertise in instructional design and, later, content strategy increased, the alignment between the disciplines began to emerge. "I don't think you can look at these different roles and put them on these linear paths without seeing the cross-section between them."

Xiong said that one of the superpowers common to both disciplines was really getting to know learners and end users by asking probing questions. "We have to truly listen to the things they are saying—and the things they are not saying—because that's really key to providing recommendations and solutions."

As her work has come to encompass both disciplines, Xiong has identified a list of common approaches shared between instructional design and content strategy that includes the following:

- Know your audience.
- Identify pain points.
- Identify goals.
- Identify knowledge (or content) gaps.
- Design a path to help close those gaps (for example, demonstrating how to get from here to there).

While alignment between disciplines is important to the creation of a sustainable content strategy practice and crucial for establishing structural integrity, it's just as important to identify points of differentiation. "During the research and conducting of interviews phase, we're trying to know our audience better, and we might ask similar questions to understand behavior or style." Both disciplines are interested in learning as much as possible about the current state of learners and users to identify problems and pain points, and both are interested in understanding their goals to help identify future state possibilities and ideas. "Did we show the user or learner that we 'get' them?"

And that's where the differentiation happens. "Our paths will then run parallel as we're designing, creating, or developing content for our respective audiences. But at the core, we're still aligned."

PROCESSES AND PLAYBOOKS

Andy Welfle, coauthor of Writing Is Designing, *Head of Content Design at Adobe*

Andy Welfle's team at Adobe does not use a codified playbook to document processes. Instead, they use case studies, which are presented internally to talk about problem spaces and show how practitioners integrate into product teams to solve those problems. "We have a pretty good culture of sharing that kind of stuff and talking about the meta-aspects of our jobs."

Because the approach to content can differ from product to product, Welfle doesn't know if a playbook would be universally relevant across his organization. Still, he is working to ensure that his team and the work they produce is well integrated into design strategy and operations. "One of the things that our design team at large is trying to do is come up with a much better articulated engagement model, and to show where the design team is positioned within the company."

Welfle has considered other ways to share information about his team—about what they do, how they work, how they add value—and to help rapidly socialize what the team is there to do. In the end, whether or not you create a playbook or socialize the work of your practice in some other format, it really comes down to determining whether this kind of documentation is necessary, who it's being created for, and what format will resonate best.

The Punch List

At first read, the concept of retooling might sound a lot like cleaning and other drudgery—necessary for keeping things in working order, but tedious and dull just the same. But retooling isn't just about tools maintenance. It's also about thinking of new and innovative ways to use the tools already available to you, as well as a chance to explore new tools that have the potential to enhance your practice:

- Whether you decide to reequip, reorganize, revise, or modify the tools used by your practice, whatever approach you take to retooling will set the practice up to scale safely and successfully.

- Consider establishing a regular cadence for retooling. Doing so primes your practice to support innovation in response to changing client needs or organizational goals.

- Creating practice roadmaps and playbooks go hand-in-hand with retooling. Producing and maintaining these documents allows you to set the course for practice growth and create a user manual of sorts to document practice operations.

Retooling, using a roadmap, and creating a playbook is about controlling the narrative and making proactive decisions about how best to scale the practice. While that may seem like a power grab of sorts, it's actually meant to reinforce two important goals: being prescriptive, in that you are positioned to reinforce content best practices; and being proactive, in that you position the practice to anticipate and solve content complications before they occur. Attention given to both of these goals elevates content to asset status—a status that's worthy of time and resources—and a status that situates the practice at the core of your agency or organization.

CHAPTER 8

Scaffolding for Sustainable Growth

In teaching, scaffolding provides support to students as they learn new concepts. Likewise, in building and construction, scaffolding is a temporary structure built to surround a building, providing support while it's being cleaned, repaired, or remodeled. The common thread in both of these scaffolding examples is the provision of support that facilitates growth. According to The University of the People, "The term scaffolding refers to a process of teaching. In scaffolding, teachers model and/or demonstrate how to solve a problem for their students. They then let the students try to solve the problem themselves by taking a step back and only giving support when needed."[1]

This chapter is a source of support for scaling, and it includes the scaffolding you'll need for the sustainable growth of your content strategy practice. The support provided here will help you assess the maturity of your practice and establish guidelines for safe scaling and growth. It will also provide some approaches you can use to manage relationships with departmental partners and others within your agency or organization as the practice continues to evolve.

In addition to introducing a practice maturity model, you'll also find adaptations of some familiar frameworks often used by many organizations—including RACI, RAPID, and SWOT—to facilitate decision-making and prioritization. If you haven't encountered these acronyms in your content career, don't fret—we'll unpack and define each one later in this chapter. You'll learn how to adapt the language of these frameworks to reach stakeholders and potential partners beyond your UX and cross-functional peers. And you'll see how doing so can position you to speak the language of leadership as you seek to cement the longevity of the practice you're building.

The Content Strategy Practice Maturity Model

In the previous chapter on retooling, you learned to assess whether or not the tools you have on hand are sufficient to handle practice expansion and how best to augment your practice toolset if the need arises. The content strategy practice maturity model looks at your readiness to scale by identifying where your practice is along a

---

1 "What Is Scaffolding in Education? Your Go-To Tips and Tricks," *UoPeople* (blog), University of the People, date unknown, www.uopeople.edu/blog/what-is-scaffolding-in-education/

continuum. It identifies growth stages by first measuring the trajectory of understanding from incompetence to competence, and also tracks the evolution from the absence of content strategy resources to the infusion of content strategy in and throughout your agency or organization. In other words, as you examine the levels and stages of maturity, you'll know where you are, so you know where (and later, how) to grow.

In an article titled "Beyond the UX Tipping Point," author and user experience expert Jared Spool wrote, "The UX Tipping Point is the moment when an organization no longer compromises on well-designed user experiences."[2] Using the introduction of the Disney MagicBand as an example, Spool pointed out that in order for Disney—or any other organization for that matter—to move to and beyond the tipping point, they must first go through a series of growth stages where user experience eventually becomes embedded into teams at every level of the organization.

NUTS AND BOLTS ACCESSING THE MAGIC OF WALT DISNEY WORLD

The Disney MagicBand is a wearable device that allows guests to interact with different touchpoints (or sensors) in the park throughout their visit. According to the Walt Disney World website, "Your MagicBand enables you to travel lighter throughout your vacation. Use it to enter the parks, unlock your Disney Resort hotel room, and buy food and merchandise."[3] The devices offer the wearer an experience that is both streamlined and personalized, and it's been cited by many UX practitioners as a great example of how to create frictionless digital experiences.

When that tipping point is reached, organizations begin shipping products that no longer compromise on quality user experiences, instead of shipping a product, feature, or service that (just) meets business requirements—and one that also has issues that will need to be fixed in subsequent releases. According to Spool, it's at this point—the tipping point—where the importance of user experience grows beyond the confines of a website or app and instead becomes entrenched in every part of an organization.

---

2   Jared M. Spool, "Beyond the UX Tipping Point," UIE, November 12, 2014, https://articles.uie.com/beyond_ux_tipping_point

3   "Understanding Magic Bands," Disneyworld, https://disneyworld.disney .go.com/faq/bands-cards/understanding-magic-band/

You may need to adjust this model to make it work for you, because every organization has unique needs and processes. Still, using a model like this to gauge practice maturity will help you scale and grow in ways that are both smart and sustainable. You'll need to be willing to take a hard look at how well your organization understands the value of content strategy, make an honest assessment of how mature your practice is (or can become), and consider how much (if at all) the value of content strategy has been (or is being) established in your agency or organization.

NOTES FROM THE JOB SITE

MY TIPPING POINT

 I attended a talk that Jared Spool gave on "Beyond the Tipping Point" in 2017. At the time, I was about two years into my role as the content strategy lead at a mid-to-large-sized organization. I was struggling to articulate to our experience design leaders how important it was for our product teams and managers to embrace the practice of content strategy as being integral to every product or service we developed for our users. I'd read Spool's article, and it struck a chord with me. So when I saw he would be giving his talk at a local meetup, I jumped at the chance to attend.

Not only was the entire talk chock-full of UX goodness, but there were also two specific concepts that, with a little adaptation, seemed to directly address some of the difficulties shared by many of my content strategy colleagues who were tasked with establishing and growing a practice: the growth stages of understanding and the growth stages of organizational UX design.

As I furiously took notes, I was excited to see how Spool's growth stages applied to the content strategy practice I was attempting to build. I saw a way that I could speak the language of our UX leaders and colleagues, using a common vocabulary to better articulate how content strategy could support the strides already being made by the organization's investment in UX. Inspired by Spool, I was able to map out a path to practice maturity—a maturity model—to demonstrate how we could scale and grow the practice to complement and eventually become embedded in UX.

Understanding Where You Are

The four stages of competence included in the maturity model were first introduced by Martin Broadwell[4] in the 1960s, and later made famous by management trainer Noel Burch. Figure 8.1 introduces the four stages from the point of view of content strategy maturity to help you contextualize Broadwell's theory.

Growth Stages of Understanding: Content Strategy at Your Organization

Unconscious Incompetence	Conscious Incompetence	Conscious Competence	Unconscious Competence
You don't know what you don't know about your organization's (or client's) content, but you know you can do better.	You know and accept that previous outcomes have added little to no value, and you continue to work toward implementing strategic approaches to content.	You continue implementing process frameworks, using existing tools and defining repeatable procedures to improve content outcomes.	You are proactively consulted on opportunities outside of product roadmaps and have created a content-focused practice roadmap.
Outcomes produced hint at adding value but miss the mark. Content strategy work comes after designs have been "baked," and the impact is minimal at best.	Outcomes are influenced by implementing best practices, such as creating frameworks and using a variety of tools to produce good outcomes.	Outcome quality improves. Content strategy is invited to product kickoffs and other product planning activities earlier in the process, and more often.	Outcomes are influenced by product and practice roadmaps. Content goals become organizational goals, as the practice delivers work that continues to demonstrate value.

FIGURE 8.1

When you move through the progressive stages of content strategy understanding, from unconscious incompetence to unconscious competence, it fosters sustainable practice growth.

In the earliest stage (the "you don't know what you don't know" stage), you likely know very little about your client's or organization's content, but you do know you can improve it. It's also unlikely that you've discovered how taking a strategic approach to content can add value, or that content should be managed as a business asset.

---

4   Martin M. Broadwell, "Teaching for Learning (XVI)," (newsletter, *The Gospel Guardian*, February 20, 1969), https://edbatista.typepad.com/files/teaching-for-learning-martin-broadwell-1969-conscious-competence-model.pdf

At this point, you're not yet able to articulate a content strategy value proposition, much less make an argument for assembling content resources into a formal practice. And that's OK. This is the learning stage—the stage where you are deep in discovery, finding out all there is to know about your content, learning the "how" of content strategy, and figuring out how to apply what you learn to existing content, perhaps making an impact on smaller projects.

It may seem that doing the critical work of discovery to get more familiar with your content will yield very little value, at least initially. Yet, as you move through the next two stages of understanding, from conscious incompetence to conscious competence, you're likely to find that you're no longer alone in your efforts to elevate the importance of content strategy. Instead, you now have a few cross-functional teammates who see the value in partnering with you, and who understand and appreciate the value you bring to the table as a content strategy practitioner.

Then, as you move toward conscious competence, these teammates will become content strategy ambassadors, advocating on your behalf to ensure that there is space at the table for the discipline and its practitioners. Without any prompting from you, they will start to ask about content—for example, pointing out that required content may "break" a new visual design—or better still, they will add you to meeting invites so you can work with your UX peers to ensure that content and design are considered together. Eventually, this process will become the norm, where content moves from being an afterthought to an anticipated (and appreciated) consideration.

Be aware that you may find yourself vacillating between these two stages, and that you may have to "rinse and repeat" with each project or client as outcomes improve. If you've implemented a process framework (see Chapter 3, "Building Materials"), keep working it—refine, adjust, and rework—as you attempt to carve out a permanent space for the practice to grow and thrive. Your work will pay off in dividends in the last stage—unconscious competence.

Now, this may seem like a misnomer at first, because why would returning to an *unconscious* state be a desirable thing? But it's here at this final stage where content strategy *becomes a given* among your cross-functional teammates, departmental partners, and other influencers in your organization. Where earlier stages demanded advocacy and awareness-building, now content considerations (and content strategy) are automatically part and parcel of every project or initiative. In some cases, treating content as an asset becomes so ingrained in the structure of the organization that you find yourself driving content-centric initiatives proactively and perhaps creating a content strategy practice roadmap where resources are dedicated to achieving content goals that add value to the business or brand.

Understanding How—and When—to Grow

In the second part of the maturity model shown in Figure 8.2, your focus shifts to content strategy maturity in your organization so that you can understand better *how and when* to scale and grow sustainably. There are five stages of maturity, beginning with situations where there's an absence of (or very few) content strategy resources, to resources being secured, to content strategy being accepted and acknowledged as a service provided to project teams on an as-needed basis, moving to the final stage of maturity, where content strategy is infused in every product team (and in some cases, throughout an organization).

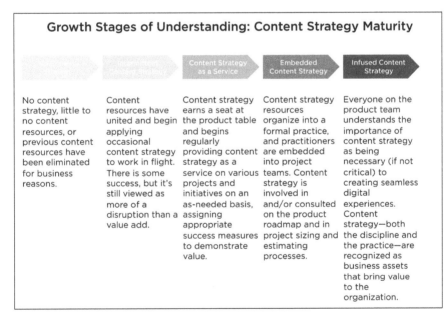

Growth Stages of Understanding: Content Strategy Maturity

		Content Strategy as a Service	Embedded Content Strategy	Infused Content Strategy
No content strategy, little to no content resources, or previous content resources have been eliminated for business reasons.	Content resources have united and begin applying occasional content strategy to work in flight. There is some success, but it's still viewed as more of a disruption than a value add.	Content strategy earns a seat at the product table and begins regularly providing content strategy as a service on various projects and initiatives on an as-needed basis, assigning appropriate success measures to demonstrate value.	Content strategy resources organize into a formal practice, and practitioners are embedded into project teams. Content strategy is involved in and/or consulted on the product roadmap and in project sizing and estimating processes.	Everyone on the product team understands the importance of content strategy as being necessary (if not critical) to creating seamless digital experiences. Content strategy—both the discipline and the practice—are recognized as business assets that bring value to the organization.

FIGURE 8.2

Charting a path to sustainable growth starts with tracking the progressive growth stages of content strategy maturity, from no content strategy to infused content strategy.

CONTENT STRATEGY MATURITY AT MCAFEE

Barnali Banerji, Design and Research Senior Manager, McAfee

Approaches to increasing the level of maturity around the value of UX content strategy may vary from one organization (or agency) to another, but regardless of where you've built your practice, it's important to start by making an honest assessment of where that level is, before you can understand how it needs to grow.

Barnali Banerji knows this all too well. "McAfee has different levels of customer experience (CX) maturity in the organization. Obviously, the highest level is within our own practice. But I don't think it's broadly understood what a content strategist can do outside of the product. Marketing content is understood, but the value of UX content is not."

To improve fluency across content disciplines, Banerji started by helping the content designers and strategists learn how to talk about their work in a different way. "They start from the user problem. Then they talk about their decisions using UX principles."

Banerji encourages content practitioners to frame their achievements in terms of how they solved business problems and to share these stories at various all-hands meetings. For example, content strategists may attend a development all-hands meeting and share how they increased development velocity.

Banerji also coaches on how to speak effectively and meaningfully to leadership. She brings a bit of role playing into the process by assuming the role of different functional leaders, such as the SVP of engineering or marketing, or the EVP of business, whose personalities she's gotten to know fairly well, having attended many meetings with them. "I try and bring a perspective that helps my team understand the things these leaders care about."

That approach also helps content practitioners learn to focus on what will resonate with a particular leader and what's important to them. "When you're able to tell stories about your success in a meaningful way, people start to understand the value that you can bring to their teams and their projects. And then they start to ask for you to help."

The growth stage's approach to scaling and growth (the maturity model) can be used in conjunction with or instead of retooling, thereby giving you another way to assess growth readiness to ensure practice sustainability, and to make sure that your practice is positioned to take on projects that will allow you to flex from a place of strength.

Methodologies for Maturity Maintenance

At the end of his talk on tipping points, Spool listed what he called the most effective plays used by design-driven organizations to maintain high levels of maturity and understanding that help the importance of good design (and UX practices overall) to remain infused and embedded at every level.

As you are pinpointing where you are in the growth stages discussed previously, it's important that you also identify ways to continually maintain and enhance practice maturity. By doing so, you avoid becoming stagnant, which can shut down pathways to scaling and growth. Here, based on Spool's plays, are a few methods you can use.

Build Content Strategy Literacy Through Immersive Exposure

You can use feedback from users to demonstrate problems faced within the digital experience that the practice can address and solve. To do this you can:

- Share mini case studies or use cases that demonstrate how content strategy can eliminate user pain points.

- Create, curate, or share relevant content strategy case studies from outside your organization to show how content strategy adds value.

- Conduct workshops to create content journey maps with cross-functional teammates and to identify touchpoints where content can address any gaps in the pursuit of a task within a digital experience.

Nurture Content Strategy Fluency Through Shared Experiences and Vision

You can provide guidance for making critical decisions through hands-on involvement. The following activities can facilitate that nurturing:

- Conduct a project-level message architecture exercise to prioritize future state content goals. Added benefits from this exercise might include:
  - Driving tactical components of content and brand communication
  - Informing a single set of brand and messaging guidelines that apply across digital experiences
- Begin (or continue) to influence organizational culture by introducing content strategy fluency at every phase of a project (where appropriate).
- Ensure that internal clients, partners, and stakeholders are reflected in content outcomes that you've collaborated on.

Achieve Shared Content Strategy Mastery by Creating a Culture of Continuous Learning

You can maintain a continual, deep understanding of business goals and user needs, always demonstrating how the practice can support both. This approach could look like:

- Always ask "what did we learn" after every major launch where the practice was involved and where content strategy made a positive impact.
- Survey internal clients, departmental partners, and stakeholders to continually assess what they know (or want to know) about content strategy.
- Create channels for continuous learning via monthly practice updates, and hosting lunch and learns, immersive workshops, and roadshows (virtual, if necessary).
- Celebrate what you've learned together with cross-functional team members and departmental partners and then rinse and repeat and do it all again.

"A message architecture is a hierarchy of communication goals that reflects a common vocabulary." —Margot Bloomstein, author, *Content Strategy at Work*[5]

Bloomstein's introduction of the message architecture as an exercise to achieve alignment on future state content goals among stakeholders, departmental partners, and other cross-functional teammates is a game-changer, especially in the context of nurturing fluency through shared experiences. Very simply put, the process involves conducting a hands-on workshop where participants work together to establish future state communication goals, using a set of cards labeled with various attributes (Bloomstein uses adjectives) to be arranged under three categories:

- Who we are.
- Who we're not.
- Who we'd like to be.

The goal is to narrow down the cards and categories to ultimately focus on "who we'd like to be."

Think of the message architecture exercise as a type of card sort, but instead of informing website navigation or structure, it fosters consensus among participants from across the organization on, as Bloomstein so eloquently put it, "what themes are most important for users to believe about the brand."

Bloomstein's book does a stellar job of spelling out the details of conducting a message architecture exercise. She's also created a pre-printed deck of BrandSort Cards[6] that you can use to facilitate your own workshop. And if they happen to be sold out (as they were at the time of this writing), grab a stack of index cards and a marker, along with Bloomstein's book, and flip to Chapter 3 for the list of adjectives she uses, or come up with a list of your own. To paraphrase Bloomstein, if you don't know what to communicate about your business or brand, how will you know where to go with your content to meet future state goals?

---

5   Margot Bloomstein, *Content Strategy at Work: Real-World Stories to Strengthen Every Interactive Project* (Waltham, MA: Morgan Kaufman, an imprint of Elsevier, 2012), 27.

6   "BrandSort Cards," Appropriate, Inc., http://cards.appropriateinc.com/

DROP IT LIKE IT'S SWOT

 Not long after I'd started writing freelance articles and blog posts as a way to get back to my reporting roots, I was offered an assignment that basically scared my socks off. ClearVoice.com is a website that connects freelance talent with brands in need of content. The site also features a robust library of articles written for and by content creators to help them with their trade.

I was approached by ClearVoice's head of content who asked me to write a piece about SWOT analysis for business and content marketing for inclusion in that library. And I kid you not, my first reaction was to hyperventilate.

This is not a good look for a content strategist who also happens to be a yoga teacher.

After putting my yoga skills to good use and doing some much-needed breath work, I was able to move myself from a state of fight, flight, and freeze to one of rest and digest. In other words, I calmed myself down, got unstuck, embraced the topic, and took some time to do a little research. And sure enough I realized that while my familiarity with SWOT was rudimentary at best, I had in fact encountered this concept once before in my agency days. I remembered that our team had conducted a SWOT analysis as part of a rebranding pitch to a potential client. Upon that recollection, my curiosity about the inner workings of SWOT was piqued just enough for me to accept the ClearVoice.com assignment.

As I did the work to understand SWOT in the context of business growth and specifically, content marketing, I thought about how in the process of being interviewed for a job, we are often asked to identify our strengths and weaknesses. I also thought about how important it is to *be prepared* (a.k.a. do the work) to answer those questions. And then I imagined how applying that level of introspection to an enterprise could make all the difference in an organization's success or failure.

I ended up cranking out nearly 2.5K words on the topic. As I wrote, I found nuggets that could translate from the content marketing applications I wrote about in the article to the UX-focused content strategy work I was doing on a full-time basis. Specifically, I considered how using the SWOT framework could help a content strategy practice establish relevance and focus on sustainable growth. Lucky for you, I made lots of notes, which have contributed to much of the information included in this section of the book.

Build High, but Safely: SWOT for Strategic Growth

For a business of any kind to survive—to be profitable and have longevity—it must grow. But growth just for the sake of growing doesn't help win the long game. Growth needs to be managed— it needs to happen strategically, and it needs to be sustainable. It also requires a business to take steps that are both meaningful and measured toward achieving long-term success.

And so it is with your content strategy practice. The premise of this book is not only about building a practice, but it's also about practice scalability and sustainability. As your practice becomes a vital part of business operations—and as content is considered and included as an asset to that business—you'll need ways to continually show how it can grow right along with (and in support of) your agency, business, or organization. There are many models and frameworks used by businesses to determine how to grow in response to changes in overall business goals and in the competitive landscape. One popular method used by many organizations is the SWOT analysis framework.

NUTS AND BOLTS THE ORIGINS OF SWOT

Albert Humphrey is credited by a number of sources as being the creator of the SWOT analysis framework during his tenure at the Stanford Research Institute (now known as SRI) in the 1960s. SWOT, which stands for *strengths, weaknesses, opportunities, and threats*, is an assessment tool to help organizations gain and maintain a competitive position in the marketplace.[7]

As a tool for strategic planning, a SWOT analysis framework is the output of an organization's honest and in-depth review of its strengths, weaknesses, opportunities, and threats, and it usually includes information similar to the bulleted lists shown under each quadrant of the generic framework example in Figure 8.3.

As the builder of a content strategy practice, you can also use the SWOT framework to inform future state goals of the practice, or to demonstrate how the practice can position your organization's (or your client's) content to overcome obstacles and support the

---

7   Richard Puyt et al., "Origins of SWOT Analysis," *Academy of Management* (July 2020), https://journals.aom.org/doi/abs/10.5465/AMBPP.2020.132

organization in the achievement of its goals. Simply put, conducting a SWOT analysis to assess where a business or organization stands in the marketplace surfaces actionable insights that can be used to inform strategies to overcome obstacles, achieve goals, and to grow safely and sustainably. By conducting that same analysis with an eye toward content as an asset, you can align your content strategy practice in a way that supports your client or organization, using a framework and language that leadership understands.

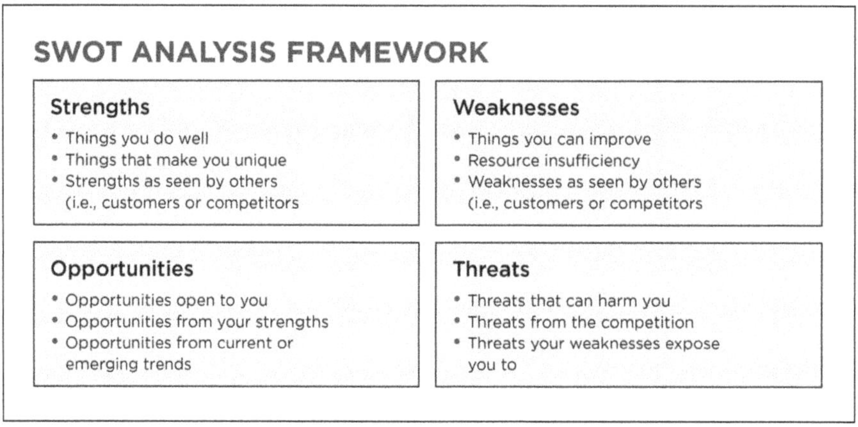

SWOT ANALYSIS FRAMEWORK

Strengths

- Things you do well
- Things that make you unique
- Strengths as seen by others (i.e., customers or competitors

Weaknesses

- Things you can improve
- Resource insufficiency
- Weaknesses as seen by others (i.e., customers or competitors

Opportunities

- Opportunities open to you
- Opportunities from your strengths
- Opportunities from current or emerging trends

Threats

- Threats that can harm you
- Threats from the competition
- Threats your weaknesses expose you to

FIGURE 8.3

A generic SWOT analysis framework can guide and document an organization's review of their strengths, weaknesses, opportunities, and threats.

Before we continue, a disclaimer: Some of the work of conducting a SWOT analysis led by content strategy practitioners gets really close (once again) to *how to do* content strategy. As you read this, keep in mind that in this example, the SWOT analysis framework (and the work you'll perform to flesh it out in support of your client or organization) isn't meant to be project-specific, though you can certainly use it that way. What you're learning here is how your *practice* can grow using the SWOT analysis framework as a tool to position the practice (and the organization's content) in a way that supports safe and sustainable scaling and growth at the organizational level.

Figure 8.4 shows an approach to creating a SWOT analysis framework from a content strategy practice point of view. You'll notice that instead of a list of statements like those shown in the generic SWOT analysis framework in Figure 8.3, this one includes a sample list of questions you might ask of client or organizational stakeholders as a way to demonstrate how your practice can strategize to ensure that there is a strategic content component applicable to each quadrant of the framework.

CONTENT STRATEGY SWOT ANALYSIS FRAMEWORK

Strengths

- Does our content clearly demonstrate what we do well?
- Does our content call out what makes us unique?
- Does our content include testimonials and other third-party content that identifies our strengths?

Weaknesses

- Do we know what content needs to be improved (where and how much?)
- Do we have sufficient resources to maintain existing and handle new content resulting from growth and expansion?
- How will content address any weaknesses that may be surfaced by others, i.e., from negative feedback, etc.?

Opportunities

- What opportunities can we support or explore by adding new content or reworking existing content?
- What strengths can we position or support as opportunities with our content?
- Are there new content trends in our industry to explore?

Threats

- Do we have a plan for creating content to address threats that can harm the organization?
- What is our content strategy for addressing threats from competitors?
- Can our content proactively address threats that our weaknesses may expose us to?

FIGURE 8.4

A modified SWOT analysis framework is used to guide and document an organization's review of its strengths, weaknesses, opportunities, and threats from a content strategy POV.

The following list of tips is based on an article I wrote for ClearVoice.com titled "SWOT Analysis for Your Business and Content Marketing."[8] It's been modified a bit here to reflect how to apply this framework as a UX-focused content strategy practitioner:

- **Maintaining success:** Ensure that you'll be able to obtain (or continue to have) the resources necessary for the practice to maintain the creation content that supports organizational goals.

- **Bring in reinforcements:** Establish a plan to temporarily augment your practice as demands for new content increase, using freelance or contract practitioners as needed.

---

8   Natalie Dunbar, "SWOT Analysis for Your Business and Content Marketing," *ClearVoice* (blog), March 23, 2020, www.clearvoice.com/blog/swot-analysis/

- **Plan for the unknown:** As new opportunities are discovered, consider whether more permanent resources—both human and technical—will be needed to support the strategic creation and use of content.

- **Expand your reach:** Explore whether it's appropriate for your client or organization to tap into additional content channels where you can expand your reach and potentially the audience for your content. Note that in some organizations, creating strategies for content used in additional channels may be handled best by marketing or similar partners, but you'll still want to include this tactic in your practice-led SWOT assessment.

- **Identify content gaps:** Look for the places where your competitors have established a presence and where your organization has not. Consider if defining an approach for creating content to address any gaps you may discover is appropriate for your organization, and if doing so will give you a competitive advantage.

- **Collaborate on the creation of an editorial calendar:** Your strengths can help you create a strategy that establishes a regular cadence for publishing the content you know is needed for your organization to maintain success. In collaboration with marketing, decide what content will be needed (and when) to shore up any weaknesses. Then plan for content needed to seize opportunities. Don't let a lack of planning turn into an internal—and avoidable—threat.

- **Invest in training:** As you anticipate market changes that may impact your business, consider (and document) the investments that will be needed to keep the practice equipped with the knowledge and skills needed to support those changes.

- **Technical obsolescence:** If your practice is responsible for the administration of content management systems and technologies, assess whether or not you have the tools necessary to ensure that the practice can continue operating in an efficient and expedient manner.

Having a framework like the one shown in Figure 8.4 at the ready—or at the very least, understanding the basic premise of SWOT and how it can support practice and organizational scaling and growth—opens up additional opportunities for your practice to operate proactively. If your client or organization hasn't yet conducted a SWOT analysis, you can introduce this framework via a content workshop or similar gathering that will get you in a room (or virtual space) with stakeholders, collaborators, and decision-makers. Getting everyone onboard with this type of assessment will inform a shared set of future state goals and establish the strategies needed to achieve those goals.

Managing Partners

The SWOT analysis framework is a great tool for defining strategies, at both the organizational and practice level. Once identified, you may need to partner with other departments to execute on those strategies, and there's a good chance that those partners will likely include marketing, as well as any legal or regulatory reviewers that your practice may be required to consult with.

A good SWOT analysis can tell you what tactics to use to support organizational growth and longevity, but it doesn't get into the weeds of who needs to execute the strategy, or who is responsible for making the decision(s) about how the strategy is implemented. For that, you can turn to two additional frameworks or models: RACI and RAPID. You can use these models to support your efforts to manage the inevitable friction points you'll encounter as the practice matures, whether from marketing partners, product teams, or regulatory (or legal) reviewers.

Another caveat: RACI and RAPID tend to be used as decision-making frameworks *at the project level*. That said, the goal of including them here is twofold:

- First, remember that like all of the scaffolding and support included in this chapter, reviews of these frameworks are provided to give you ways to communicate and collaborate with various teammates and departmental partners using a language they know and understand.

- Second, while generally project-focused, with a little tweaking and imagination, you can adapt these frameworks to use as a way to define how the practice will interface with departmental partners who may be outside of your immediate cross-functional team.

Just as with SWOT, you can create visual representations of RACI and RAPID that can be easily accessed and referred to for those (inevitable) times where there is friction between the practice and others on who does what and when. Things can get messy when departmental partners feel they're being "told what to do" by those outside of their team (for example, you or your practitioners). As well, there may be processes internal to other teams or departments that create dependencies that sometimes manifest as delays to the timely execution of the strategies your practice creates, or even the content that is created by others. Making time to sit with your departmental partners to work together on documenting interactions proactively with the practice (and the processes that support those interactions) will go a long way toward anticipating those inevitable growth pains and managing partner participation and expectations.

RACI: Responsible, Accountable, Consulted, Informed

Here's a question that many content strategists have likely heard in one format or another that's stopped a project dead in its tracks: who is responsible for obtaining approvals on content created for a digital experience? Is it the folks who write the copy and curate content? Or is it the strategists who recommend the content to create and curate in the first place? Or, to really muck up the works, is it both?

If that's not enough of a quandary, consider these questions:

- Who is *responsible* for identifying outdated content?
- Who is *accountable* for removing outdated content?
- Who should be *consulted* on decisions about removing content?
- Who needs to be *informed* when content is removed?

As a decision-making tool, the RACI model clearly defines roles and responsibilities (see Figure 8.5). As you are building your practice, creating a RACI chart can help you visualize decisions reached in collaboration with other departments and team members, and it can also help you further align on practice operations. Simply put, a RACI chart defines who is who, and what it is that they do, all from a content strategy practice point of view. When you strip down the questions in the example at the beginning of this section, a RACI matrix asks and answers four basic but very important questions that are necessary for getting work done:

- Who is responsible?
- Who is accountable?
- Who should be consulted?
- Who should be informed?

The interactions between and responsibility of content teams from marketing and UX are clearly spelled out in the RACI matrix example in Figure 8.5. It's particularly helpful when both teams are involved in creating and supporting the end-to-end experience of a product or feature within a digital experience.

RACI: GETTING THE WORK DONE

PROJECTS/TASKS	DEPARTMENT/TEAM	
	USER EXPERIENCE	MARKETING COMMUNICATIONS
PRODUCT OR FEATURE OWNERSHIP	Accountable Responsible	Consulted
VISUAL DESIGN DELIVERABLES	Accountable Responsible	Consulted
CONTENT STRATEGY DELIVERABLES	Accountable Responsible	Consulted
CONTENT CREATION AND DELIVERY (TRANSACTIONAL)	Consulted	Accountable Responsible
CONTENT CREATION AND DELIVERY (MARKETING)	Accountable Responsible	Consulted
INTEGRATION OF CONTENT AND VISUAL DESIGN	Accountable Responsible	Consulted
BRAND MESSAGING	Consulted	Accountable Responsible

FIGURE 8.5
This simple RACI matrix shows how to facilitate and document how the content work gets done when there are content partners from UX and marketing.

As a tool for building and cementing your content strategy practice, a RACI matrix will define how you'll work with your departmental partners to get the content work done.

NUTS AND BOLTS THE ORIGINS OF RACI

The RACI model has been in use since the 1950s and was first known as the *Decision Rights Matrix*, according to some sources. Other sources indicate that the tool originated from Goal Directed Project Management (GDPM), created in the early 1970s. A book of the same name, authored by Erling S. Andersen, Kristoffer V. Grude and Tor Haug, was first published in 1987. There are four editions of the book, with the latest being published in 2009.[9, 10]

---

9 "What Is RACI? An Introduction," *RACI Solutions* (blog), www.racisolutions.com/blog/what-is-raci-an-introduction

10 Erling S. Andersen et al., *Goal Directed Project Management*, 4th ed. (London, UK: Kogan Page, 2009), www.koganpage.com/product/goal-directed-project-management-9780749453343

RAPID: Recommend, Agree, Perform, Input, Decide

Let's return to the scenario from the previous section on RACI and consider the questions asked about who is responsible for obtaining content approvals. Let's say that the responsibility for that activity has been given to your marketing partners, and that if needed, your practice provides consultation on the underlying strategy that informed the content marketing and is seeking to have it approved. There are likely underlying steps that, when taken together, define that approval process. RAPID unpacks what its creators, Bain & Company, call decision accountability.[11]

The questions that RAPID seeks to answer are:

- Who recommends an action to take?
- Who needs to agree with that recommendation or action?
- Who provides input into the recommendations made?
- Who makes the final decision (on what action to take)?
- Who executes (or *performs*) the actions in support of the decision?

A simple search on the internet will result in myriad ways you can visually represent RAPID in a diagram. However, the diagram created by Bain & Company is the simplest and most easily digestible. Figure 8.6 is based on that diagram and shows the five steps involved in the process.

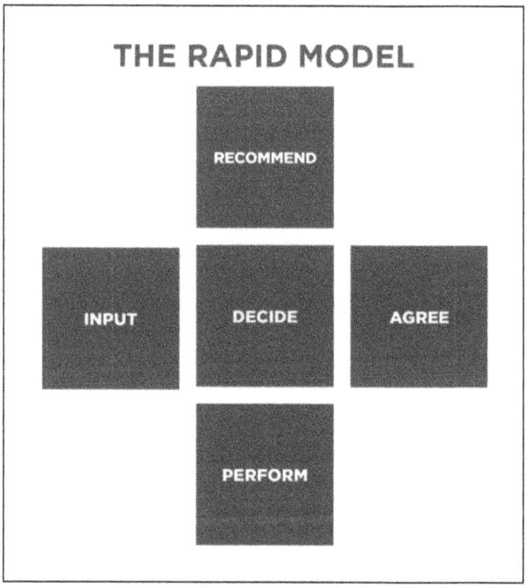

THE RAPID MODEL

RECOMMEND

INPUT · DECIDE · AGREE

PERFORM

FIGURE 8.6
You can use the basic elements of the RAPID matrix—recommend, agree, perform, input and decide—to define decision accountability that involves content.

---

11 "RAPID®: Bain's tool to clarify decision accountability," Bain & Co., August 11, 2011, www.bain.com/insights/rapid-tool-to-clarify-decision-accountability

While every part of the RAPID model is important, all of the elements on the periphery are key to support the person ultimately responsible for making informed decisions.

Many organizations will create a matrix as a visual representation of the RAPID process. This matrix involves more detail than the previous diagram and provides details about tasks that are required and that need to be executed once a decision is made.

As with the RACI model shared in the previous section, the interactions between and responsibility of content teams from user experience and marketing communication are clearly spelled out, in the matrix shown in Figure 8.7, this time with the goal of determining who is responsible for making decisions about what content work gets done and how it gets done. Note that you can also create a RAPID matrix that includes other disciplines from your cross-functional team.

RAPID: MAKING DECISIONS

PROJECTS/TASKS	DEPARTMENT/TEAM	
	USER EXPERIENCE	MARKETING COMMUNICATIONS
PRODUCT OR FEATURE OWNERSHIP	Recommend	Agree
VISUAL DESIGN DELIVERABLES	Recommend	Input
CONTENT STRATEGY DELIVERABLES	Recommend	Input
CONTENT CREATION AND DELIVERY (TRANSACTIONAL)	Recommend Perform	Input
CONTENT CREATION AND DELIVERY (MARKETING)	Recommend Perform	Input
INTEGRATION OF CONTENT AND VISUAL DESIGN	Recommend Perform	Input
BRAND MESSAGING	Input	Recommend

FIGURE 8.7
This example of a RAPID matrix documents how *decisions* about the content work are made, again in an organization where both UX and marketing may have overlapping content roles.

In the context of building a content strategy practice, the RAPID model details how content strategy decisions are made. The model may include how content changes are initiated and communicated, as well as the standards (policies) and oversight (people) that guide content implementation.

CONTENDING FOR CONTENT STRATEGY

When it comes to defining what the discipline content strategy is, where the practice belongs within an organization, and who is responsible for executing the work, absent alignment and collaboration, there is no clear winner. I've found this to be especially true where there is one content strategy practice rooted in user experience and a separate practice in marketing.

For there to be a seamless experience for users, the process for the creation of content that supports that experience should be as seamless as possible, too. Friction between teams can introduce friction within an experience. When that happens—when there is contention among the teams responsible for creating a singular experience, your audience—the users of your products and services—don't get the benefit of a best-in-class user experience. In fact, as I've witnessed during research sessions, savvy users can sometimes detect a disconnect in a digital experience.

Rather than debating who "wins" the coveted title of content strategy, the better approach (as discussed ad nauseam throughout this book) is to align first on the work that needs to be done, and in doing so, focus on a goal that is shared. Because in the end, when the fighting stops and there is still no alignment, it's often the customer who loses out.

IDENTIFYING THE DECIDERS

While the decision-making frameworks covered in this chapter can facilitate conversations around roles and responsibilities, there are times where the work is too ambiguous to have a clear owner or approver. Or as a wise content strategist once put it, sometimes figuring out "who decides who decides" is not easily solved within a framework.

In my experience, these models have definitely helped. But other times, I've scrapped the model completely in favor of just having a one-on-one conversation with a marketing partner, for example, and have often had some success once the number of voices in the room was reduced. For the few times that approach hasn't work, the matter gets escalated to departmental leaders to hash out.

RAPID was created by global consultancy Bain & Company and is described as "a tool to clarify decision accountability." RAPID is, according to Bain, "a loose acronym" for recommend, agree, perform, input, and decide. "RAPID® assigns owners to the five key roles in any decision."[12]

The Punch List

So this chapter ends where it began—with scaffolding. Scaffolding offers the support you need to build a sustainable practice by introducing a variety of approaches (or layers of assistance) that measure practice maturity, identify roles and responsibilities, and support informed decision-making—tactics that will eventually become second nature as you continually evaluate opportunities for scaling and growth.

This chapter opened with a practice maturity model you can adapt and use to measure organizational readiness as you consider practice expansion, scaling, and growth.

You can also use the following decision-making frameworks (models or matrices) to assess organizational content goals, establish how content work gets done, and to identify those responsible for making content decisions, including the following:

- SWOT, which provides an honest assessment of an organization (or your practice) that gives actionable insights to help overcome obstacles and achieve goals
- RACI, which addresses how work gets done by clearly identifying roles and responsibilities, and can be utilized within your practice, or between the practice and other collaboration partners
- RAPID, which is a tool that identifies and then supports those people responsible for making informed decisions about critical actions to be taken toward the achievement of future state goals

You can turn to these tools as you nurture, maintain, and scale your practice to meet future state goals and market demand for your clients or for your organization.

---

12 "RAPID®: Bain's tool to clarify decision accountability," Bain & Company, August 2011, www.bain.com/insights/rapid-tool-to-clarify-decision-accountability/

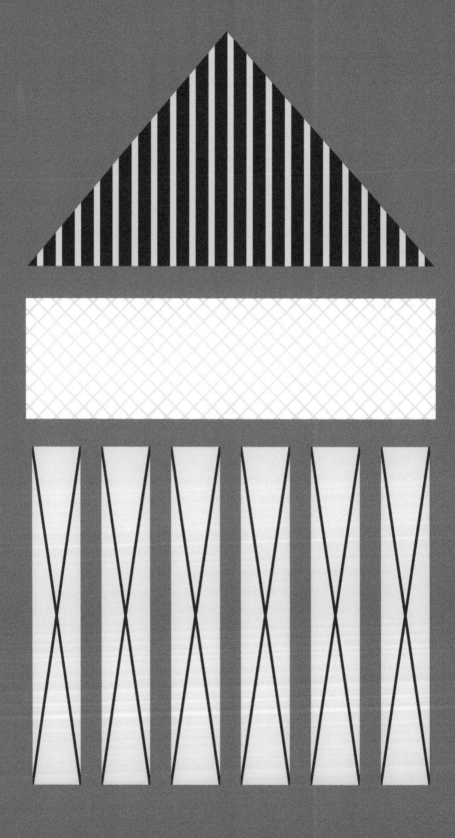

CHAPTER 9

Inspection-Ready: Engaging the Executive Suite

In Chapter 2, "Structural Alignment," you read about the importance of alignment through the establishment of relationships and creating alliances with cross-functional teammates, departmental partners, and anyone else in your agency or organization who had a vested interest in the success of the practice you've built. You also learned how important it is to continually nurture those relationships and partnerships in support of practice sustainability and longevity.

Now, using those same relationship-building skills as a foundation, you're going to elevate the exposure of your practice by demonstrating its value to stakeholders and key organizational leaders, sharing use cases, lessons learned, and measurable outcomes all the way up to the C-suite. For example, you may have created a process that works in an Agile environment to ensure that content is ready to ship on time (thus avoiding potentially costly delays). Or perhaps you've developed end-to-end strategic content solutions that have resulted in the creation or curation of content that has impacted important engagement metrics positively. Whatever strides you've made, you'll soon be able to demonstrate the value that a solid content strategy practice can have at the appropriate levels of leadership across the organization—and you'll understand why it's vitally important to do so.

Phases of Inspection

In the building and construction trades, there are several milestone phases where the inspection of critical structural elements happens. For example, look at when the foundation is poured; when essential systems are installed (think: plumbing and electrical); when exterior structures are built (think: the walls and roof); and of course, there's the final inspection, where building inspectors, developers, and other invested parties sign off on the integrity of the structure, deeming it ready—and safe— to occupy.

Likewise, as you work to figure out *how* to engage leadership in the workings of the practice, you'll also discover *when* it's best to do so. An example of when engagement might be appropriate is during quarterly reporting or during an all-hands meeting where you can highlight practice achievements.

Another proactive way to address when to engage leadership is to schedule progress check-ins, where you can share significant practice milestones. More specifically, it could be through the sharing of use cases and other examples of how the practice addressed an identified problem that was encountered regularly by users, perhaps creating a content-led solution. Or it could be how the practice repositioned content that satisfied a previously unmet user need. This approach pulls from project-level examples and can be most effective in an environment where the seeds for building the practice were planted at the departmental level.

Self-organized practitioners may find it beneficial and necessary to engage the C-suite as well, perhaps to highlight progress, gain alignment, or request funding to grow practice operations and capabilities. If this sounds like your situation, you'll also want to discover ways to be proactive about engaging leadership. Try conducting a message architecture exercise (shared in the previous chapter), or consider involving leadership in content-led design sprints (covered later in this chapter) as a way to demonstrate practice prowess.

Why Leadership? Why Now?

There's a chance that you may be wondering about the logic behind using relationship-building skills in the context of presenting the practice to leadership. You may also be wondering why this logic is being presented at the end of the book, *after* you've learned about the blueprint components, along with retooling, scaffolding, and related techniques that stabilize and scale the practice.

Or to put it another way: Why leadership, why now, and why not before introducing the blueprint components? Let's address the timing of information question first—why now and not before? There are myriad variables to consider when deciding *when* it's best to engage leadership in the development of the practice, with the most significant one being organizational structure, followed by the processes and procedures that are in place in your organization. Here's a look at a few variations of how structure and process can work in an agency or company, and some suggestions for how, when, and why a practice lead might use or introduce the Content Strategy Practice Blueprint to leadership—and if necessary, the request to build a practice:

- **Traditional, top-down organizations:** Start with this chapter and learn the importance of positioning the creation of the practice as an integral part of overarching business goals if you work in an organization where you are required to run recommendations and changes in operational processes up the proverbial flag

ACTIVE LISTENING

Berni Xiong, author, coach, content strategist, and instructional designer

Active listening helps you navigate crucial conversations with ease, whether you are speaking with cross-functional teammates and departmental partners, or presenting the value of your content strategy practice to leaders and stakeholders.

Active listening was first defined as a communication technique by Carl R. Rogers and Richard E. Farson. "There are many kinds of listening skills [but] the kind of listening we have in mind is called 'active listening.' It is called 'active' because the listener has a very definite responsibility. [They] do not passively absorb the words which are spoken to them. [They] actively try to grasp the facts and feelings in what [they] hear, and [they] try, by their listening, to help the speaker work out their own problems."[1]

Industrial designer Berni Xiong takes the concept a step further, combining active listening with curiosity and empathy—abbreviated ACE—to garner feedback, building on a technique she learned from authentic marketing coach, George Kao. Figure 9.1 shows how these components work together when gathering feedback.

"When you're looking to get feedback from your customers or users, you literally ask them for their words—you ask them for keywords, you ask them

---

1 Richard E. Farson and Carl R. Rogers, *Active Listening* (Connecticut: Martino Publishing, 2015).

pole. You'll also share the blueprint components with leadership, so they are aware of and aligned with the approach.

- **Autonomous organizations:** Work through the blueprint (and this book) from the beginning if you work in an organization where leaders at the operational level, such as the directors or senior managers of a UX team, are entrusted with decision-making. You'll still want to provide leadership with updates at key points in the building process and work toward establishing a regular cadence for reporting results as you build, and later, as you scale the practice.

for relevant terms. Getting into the mind and heart of the user, learner, or customer helps you to enhance their experience," Xiong said.

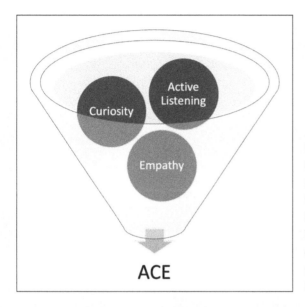

FIGURE 9.1
Combining active listening with curiosity and empathy, ACE instills confidence in leadership and demonstrates how your practice will achieve business goals.

As a communication technique, ACE works well when you're demonstrating the value that your practice can have to leadership and across the organization. According to Xiong, using ACE lets you get to know your audience better—to understand their problems and goals. "Ask people what they want. Then use their language in your messaging to tell them what they want to hear."

- **Self-organizing content (or UX) team:** Start by working the practice blueprint at a smaller scale. Gather a few use cases that you can share with your immediate managers or directors as a way to secure their buy-in and learn the best way to engage leadership when it becomes necessary, such as when funding is needed or when you're ready to scale the practice and need to add headcount or secure new tools to help the practice scale successfully.

A (DESIGN) SPRINT TO THE FINISH

In Agile, a sprint is usually a time-boxed period (often two weeks, but it can be 10 days or even up to four weeks) where a team works to finish a predefined amount of work to create or iterate on a new product or feature. The beauty of sprints (yes, I am a big fan) is that they allow you to do a set amount of work, present it to your cross-functional team for review and inspection, and glean learnings from that work that inform the next set of tasks to be completed as you iterate toward a final solution.

And so it is with design sprints, a five-day design process first introduced at Google Ventures by Jake Knapp.[2] The goal of a design sprint is to get laser-focused on a problem you'd like to solve—without distractions, meetings, or interruptions—and to methodically work through a five-phase, five-day process that includes:

1. Understanding the problem space
2. Sketching out and ideating on potential solutions
3. Deciding on which solution is the best to move forward with
4. Prototyping the winning solution
5. Validating the solution for real world potential

So what does all this have to do with your content strategy practice? And how does a design sprint establish a bridge for communicating with leadership? First, design sprints by their nature are meant to be collaborative and cross-functional. Second, content strategy is a perfect opportunity to shine during these tactical, time-boxed missions. When you remove *lorem ipsum* or placeholder copy from prototypes, and instead have a content strategist not only create copy, but also validate flows and functionality, you tell a story from the user's point of view that leaders of the business can see, relate to,

---

2 "The Design Sprint," Google Ventures, www.gv.com/sprint/

The previous scenarios represent just three of many possible ways an organization might be structured, In addition to providing approaches for engaging the C-suite, these scenarios also address when it's best to do so. Because every organization is different, you'll need to do the legwork to decide what works best for engaging the leaders of your agency or enterprise. Again, you can refer to the

and understand. You also avoid what happens all too often in the real world of product development: bringing in content too late in the process and expecting copy to be created with little to no context for the problem space, or worse—not fully understanding the user problems you're trying to solve.

I've been lucky enough to participate in similar processes that predated the 2010 introduction of design sprints. Those early explorations were usually led by engineers, and were inclusive of visual designers who, somewhere along the way decided it would be good to have a content person in the room—and sometimes, the content person was the one *leading* the challenge. An example of this was shared in Chapter 3, "Building Materials," when I was tasked with finding a solution for placing promotional content on the home page of the large online directory site. I pulled together a cross-functional team that included marketing, visual design, and engineering. We threw a bunch of sketches (and Post-it Notes) on the walls of many conference rooms, and we imagined what might be possible beyond the existing experience. Eventually, after discussing feasibility and figuring out what the experience might look like (and how it would function), we presented our ideas to leadership.

During my tenure at another company, I was part of a cross-functional team that participated in a remote design sprint that lasted over six days or so. The hours were long (especially over Zoom), and often we found ourselves working outside of our specific disciplines to meet critical deadlines. In the end, our cross-functional efforts paid off in a big way: our team was one of four challengers (out of a field of over 80) awarded a share of a five-year, $247 million dollar contract to contribute to the ongoing improvement of customer experiences on of one of the largest government web properties in the U.S.

So whether you're participating in or driving a design sprint or similar activity, you'll demonstrate the importance of getting the practice involved in digital experiences and product development from the time of kickoff. And you'll show why content strategy should *not* be an afterthought in both design-focused sprints and in real-world product development.

methods for building alliances outlined in Chapter 2, to help with this process, including defining the space and scope of the practice, aligning the practice purpose to support strategic business outcomes, and creating a shared advocacy for content as an asset to be valued by and across the business.

At this point, you may be wondering, if leadership engagement isn't required in your organization—that is, if you work in a company or agency that is on the more autonomous end of the leadership engagement scale—do you really need to know the information covered here? The answer is a resounding *yes*! Your role as a leader and practice-builder here is pivotal. The very health and longevity of the practice depends on leadership's willingness to embrace and support what was discussed in Chapter 8, "Scaffolding for Sustainable Growth"—building literacy and content strategy competency *at every level of the organization.*

It's not just about approaching key leaders for funding or seeking a larger budget. Sure, these things are important, and it's hard to scale up or out without sufficient financial resources. But it's also about doing the work necessary to actualize a culture of continuous learning and foster content maturity. As business goals and user needs adapt to a changing marketplace, leadership will not only see the value your practice brings, but they'll also understand that the services the practice provides are integral to strategic planning in support of future state organizational goals. Even if your organization is autonomous and you don't need leadership's permission to grow your practice, you'll still want them to understand why the work of the practice is important to the goals of the organization as a whole. Fostering that understanding can shift the company culture in a positive way toward viewing content as an integral business asset.

Decoding Business Speak

So now that you have an idea of when, where, and how to engage business leaders, let's unpack the next level of *how* and learn to master the language of the business leaders you want to communicate with. To grab the attention of leadership, you'll need to master the language of business speak. To do that effectively, you'll need to unravel the code—usually in the form of acronyms—often spoken by organizational leaders and stakeholders. This exercise may feel counterintuitive to your content strategy sensibilities, because it flies in the face of one of the discipline's main best practices—that of not using jargon. But in this case, making an exception to that rule will help you articulate how the practice aligns and scales with the goals of the business in a way that resonates with those in charge.

Before the decoding begins, it's important to remember that every business uses its own vocabulary to communicate things like success measures, metrics, and similar concepts, and it's vitally important that you take the time to discover which acronyms and related terms are important to your leadership team and to understand the context within which they speak about them. Reading quarterly reports, attending all-hands meetings, and setting up short internal interviews with key stakeholders are just some of the ways that you can begin to understand the language of leadership in your agency or organization. You can also check with your user experience partners—visual designers, human factors engineers, design researchers, and others—to see if they've done similar work, and if there is useful information they can share. As more and more agencies and organizations invest in building multidisciplinary design teams, those teams are being tasked with showing the ROI (return on investment) that the company will reap from making those investments. (And yes—ROI is one of the many terms you'll need to know and understand when communicating up the organizational ladder.)

Please note that as you review the terms listed in Table 9.1, you won't find instructions on how to calculate the metrics or measures included here. There are myriad how-to books available that delve deeper into metrics calculations and applications. Better still, seek out opportunities to establish an information exchange with those closest to monitoring metrics in your organization. Not only will this exchange help you with your goal of becoming more fluent in the language of leadership, but it will also have the added bonus of creating alliances with additional disciplines as you're building the practice.

One more thing to note: Some of the measurements included here have their origins in marketing. And while the focus on this book is on building a *UX-focused* content strategy practice, it's likely that your work either influences or overlaps with your organization's marketing efforts, and that marketing is (or will soon become) one of your primary departmental partners. So in addition to leveling up on leadership lingo, knowing and understanding how to use these (and similar) terms will also go far toward strengthening the relationship between you and your marketing partners.

TABLE 9.1 DECODING THE LANGUAGE OF LEADERSHIP

Acronym	What It Stands For	What It Is	Content Strategy Context
AOV	Average Order Value	This measure is applicable to websites and apps that are in the business of selling products or services, and it looks at trends in the average amount that customers spend, divided by the company's total revenue.[3]	Content strategy practices focused on creating product-specific content can use AOV as an input to improve the experience that users have with a product, which can contribute to a higher AOV. If AOV starts to trend lower, a product content strategy practitioner can do the work to better understand whether or not product content is delivering the best possible experience, enabling users to interact with the experiences created by the product.
CSAT	Customer Satisfaction	This measure looks at whether or not a customer (or user) is able to achieve a level of "completeness" when engaging with your digital experience, products, or services. It's not just about making customers happy, or whether or not they've enjoyed the experience you've created; it's more about the ability to complete a task, such as making a purchase or finding needed information.[4]	An overall best practice of content strategy is to ensure that users get the right content at the right time and in the right format or channel. The content strategy practice can ensure that there are methods to develop the systematic implementation of this approach across the enterprise, beginning with the establishment of a core strategy via a message architecture exercise or similar methodology.
CX	Customer Experience	This measure considers how customers value or perceive your brand overall. Your team may influence some of the many touchpoints that your customers encounter when engaging with your brand, but the end result of that work—brand perception—rests with the customer.[5]	It's difficult at best to assign a specific score to a person's perception. Loosely put, perception comes down to this: your brand is what customers say it is, not what you say it is. But how—and where—content is used to influence what customers think about your brand falls within the practice's sphere of influence, as does the creation and curation of content that supports how users experience your brand via digital interactions.
KPI	Key Performance Indicators	As a collection of measurements, KPIs track an organization's progress toward a set of organizational goals. KPIs are quantifiable: once goals (or targets) are set, you use KPIs to measure how the progress is made toward that goal.[6]	The following short list includes examples of KPIs that content practitioners may want to track: who consumes the content created by your organization (audience segmentation); how many of those consumers are repeat users (retention): and how do these users engage with the content (engagement). These KPIs are good indicators for showing where the practice can have direct impact.

continues

---

3 "Optimization Glossary," Optimizely, www.optimizely.com/optimization-glossary/average-order-value/

4 "Your Ultimate Guide to Customer Satisfaction," Qualtrics, www.qualtrics.com/experience-management/customer/customer-satisfaction/

5 "What Is CX? Your Ultimate Guide to Customer Experience," Qualtrics, www.qualtrics.com/experience-management/customer/customer-experience/

6 "What Is a Key Performance Indicator (KPI)?," KPI.org, https://kpi.org/KPI-Basics

TABLE 9.1 CONTINUED

Acronym	What It Stands For	What It Is	Content Strategy Context
NPS	Net Promoter Score	Have you ever received a single-question survey after using a website or otherwise engaging with a brand that asks something along the lines of, "How likely are you to recommend our brand?" That question is used to glean a Net Promoter Score, which is an indicator of how likely a customer is to recommend—or "promote"— your brand to others. According to NPS creators Bain & Co., "[NPS] is a single, easy-to-understand metric that predicts overall company growth and customer lifetime value."[7]	You can use trends in NPS scores to dig deeper to find if there are strategic content solutions that can help improve lower scores or help maintain scores at the higher end of the range. Most times, the "how likely are you to recommend" inquiry is actually just the first of several questions in an extended survey, where participants are sometimes asked to explain the scores they gave. That type of qualitative information is a potential goldmine for content strategy practitioners, because it may include specifics that can guide content decisions that will impact (and ultimately improve) the experience of your users.
OKR	Objectives and Key Results	Most sources credit Andy Grove with the creation of OKRs—Objectives and Key Results—while he was at Intel. Simply put, objectives are the things the business wants to accomplish. They are "significant, action-oriented, and inspirational." And key results are how the company will accomplish these objectives. They are "specific, timebound, aggressive yet realistic, and verifiable.[8]	Your practice can either support organizational OKRs or create practice specific OKRs that will have an organizational impact. The practice might support UX teammates in a goal of creating prototypes by a certain date as an organizational KPI. On the other hand, a practice-focused KPI might include creating or updating a style guide, contributing content guidelines and assets in a design system, or tracking internal engagement in practice-led lunch-and-learns, with a goal of increasing awareness and attendance over time.
ROI	Return on Investment	Simply put, as a performance measurement, ROI assesses how profitable a business investment is. According to the Nielsen Norman Group, "Calculating ROI is a *powerful tool for building buy-in*, because it can demonstrate that UX isn't just good for users—it's also very good for the business."[9]	As a content strategy practitioner, you want to demonstrate how content can contribute to predefined actions that you want your users to take, which are deemed important to the business and its overall goals. Depending on your industry or the types of products or services your company provides, some of these actions might include buying a product, paying for a service, or signing up for a subscription.[10]

7   "Net Promoter System," Bain & Company, Bain.com, www.bain.com/consulting-services/customer-strategy-and-marketing/customer-loyalty/

8   Ryan Panchadsaram, "What Is an OKR? Definitions and Examples," What Matters, www.whatmatters.com/faqs/okr-meaning-definition-example/

9   Kate Moran, "Three Myths About Calculating the ROI of UX," Nielsen Norman Group, September 6, 2020, www.nngroup.com/articles/three-myths-roi-ux/

10   Jason Fernando, "Return on Investment," Investopedia, September 13, 2021, www.investopedia.com/terms/r/returnoninvestment.asp

The previous list of terms is by no means exhaustive. Nor is it meant to represent which terms are "right," or which are the "most popular." The terms used by the leaders in your organization might differ from those listed here. In fact, it's a safe bet that by the time this book is printed, this list might include some new entries, while others listed here may be deemed less important. Whenever this book happens to land in your hands, you'll need to do your due diligence to learn what terms hold the most significance among the key leaders of your agency or enterprise.

POWER TOOLS RESOURCES ON METRICS AND ANALYTICS

If the idea of anything having to do with numbers makes your content-focused brain hurt a little bit, I can relate. I got into words so I could stay far, far away from anything remotely requiring math. But in today's data-driven digital landscape, there's no escaping the importance of metrics and analytics.

While it's a very good idea to get to know the people in your agency or organization whose primary focus is to make sense of things like abandonment rates, conversion rates, impressions and more, there are some great resources out there to help you dive deeper into all things data. Here's a short list to get you started:

- *Measuring the User Experience: Collecting, Analyzing, and Presenting Usability Metrics (Interactive Technologies)* (3rd Edition) by Tom Tullis and Bill Albert
- *Practical Web Analytics for User Experience: How Analytics Can Help You Understand Your Users* by Michael Beasley
- *Search Analytics for Your Site: Conversations with Customers* by Louis Rosenfeld
- *Web Analytics, An Hour a Day* by Avinash Kaushik

Change Orders: Meeting Marketplace and User Demands

Another pivotal conversation you'll want to have with leadership involves how the practice helps the organization adapt to business goals and user needs in response to changes in the marketplace, changes in the brand (including changes in products and services offered), or changes in the needs of users. The SWOT analysis introduced in the previous chapter (shown again in Figure 9.2) is a great tool you can use to help open communication channels with

WHEN THE CEO MET SEO

During my formative years in digital content, circa 2006, search engine optimization (SEO) was all the rage. From conducting keyword research to discover the words users searched for the most, to leveraging those keywords for maximizing top placement and increasing traffic, everyone who ran a website for any enterprise wanted a piece of the SEO pie.

I can't say with any certainty whether the CEO of the online directory company where I worked was instrumental in bringing in a product manager whose only focus was SEO. Nor do I know how that CEO came to embrace the importance of SEO (and later, metrics like unique visits, page views, and so on). But one thing I can say for certain: once we'd hired that SEO-focused PM—and once our content team sat and strategized with that person on how we could strategically use content to improve things like our overall position in search results or increasing the number of unique visits to our site, rarely a week (and maybe even a day) went by when that CEO would walk by our cluster of cubicles and ask, "What's our ranking looking like today?"

Never mind the fact that to gain any real traction after implementing content changes and other strategies, it usually took months for those changes to impact page rank and other metrics.

As part of our strategy, we created editorial features that increased repeat visits and led initiatives to update and revise the design and content for localized city guide pages. The results of our work earned the respect of our CEO and other leaders from across the organization: we increased SEO by 40 percent and caused a lift in our page rank within the first three months after launching the strategy. We also received funding to secure much-needed tools to push content live, independent of major code releases. Ultimately, we improved the metrics relevant to leadership across the board: we provided users with content that was fresh, timely, localized, and relevant, and we did it all by learning how to speak the language of leadership.

While the constant inquiries could be annoying, because the work we did garnered such amazing results, content was invited to have a literal seat at the table to participate in the decision-making process for vetting and selecting analytics tools. Because they were able to see and quantify the contributions of our content team, the CEO became more interested in and engaged with the work we were doing and how it contributed to the overall goals of the business.

leadership. As you learned previously, SWOT is a strategic planning tool that gives an organization actionable insight via the output of an analysis of an organization's strengths, weaknesses, opportunities, and threats.

SWOT ANALYSIS FRAMEWORK

Strengths
- Things you do well
- Things that make you unique
- Strengths as seen by others (i.e., customers or competitors

Weaknesses
- Things you can improve
- Resource insufficiency
- Weaknesses as seen by others (i.e., customers or competitors

Opportunities
- Opportunities open to you
- Opportunities from your strengths
- Opportunities from current or emerging trends

Threats
- Threats that can harm you
- Threats from the competition
- Threats your weaknesses expose you to

FIGURE 9.2

The SWOT analysis framework can be used to guide and document an organization's review of its strengths, weaknesses, opportunities, and threats.

Let's revisit each quadrant of the SWOT matrix shown in Figure 9.2 to discover more ways that you can use this assessment tool to demonstrate the value of the practice to the leaders in your agency or organization from a practice point of view.

- **Strengths:** Leaders tasked with identifying organizational strengths seek to identify what the business does well: what gives the business an advantage over competitors, and what is uniquely different about your business when compared to similar organizations. From a practice point of view, it's imperative that leadership continues to make an investment in keeping content updated to ensure accuracy and that content is used properly to communicate those strengths in appropriate formats and channels.

- **Weaknesses:** The hallmarks of identifying weaknesses are honesty and objectivity. This process requires organizational leaders to assess areas for enhancement and change in order to improve relevant business results, such as more sales, an increased number of visitors or users, and so on. From a practice point of view, you can help leadership understand where there may be gaps in content that, if left unaddressed, can lead to the exploitation of weaknesses, causing your target audience to look to other organizations for what they need.

- **Opportunities:** When considering opportunities, leaders want to identify trends or other changes in the marketplace that have the potential to grow their organization's reach. From a practice point of view, you can help leadership understand how the practice reviews and includes inputs from various sources, such as comparative analysis, user research, and other business intelligence to inform future state content goals in support of future business opportunities.

- **Threats:** When identifying threats, leaders look to factors outside of the control of the organization that can put the business at risk, such as changes in regulations that are applicable to the business. A thorough threat assessment should also consider the root cause(s) of potential threats so that the business can identify ways to respond thoughtfully, instead of being reactionary. From a practice point of view, establishing a content response plan proactively to potential threats can better prepare the organization when threats occur. It's not possible to completely be prepared for the unknown (think COVID-19), but it is quite possible to establish the pivotal role of content—and by extension, the practice—in crisis mitigation.

The ultimate goal of conducting a SWOT analysis is to provide information to an organization that is actionable and provides direction for establishing strategies to achieve its goals. When engaging leadership, you can use this same methodology to show how the practice can and should be positioned as an integral part of the SWOT analysis process at the organizational level.

EFFECTIVE ENGAGEMENT

Kristina Halvorson, founder and CEO of Brain Traffic, and author of Content Strategy for the Web

Kristina Halvorson has decades of experience working with companies of all sizes. So she is quite knowledgeable about when and how to engage executives, as well as whether it's even necessary to do so. "The first thing you need to do is get whoever you're doing the good work for to go to *their* boss, or to their boss's boss. That's laying the groundwork for sponsorship internally and building your fan base."

Next, it's important to identify the right audience—for example, stakeholders, product managers, or design leads—and then figure out what matters to them. "You have to talk to them about what they care about or talk to them about content strategy *through the lens of* what they care about."

Some of those concerns might be framed in questions that sound like, "How can you help me reclaim lost revenue? How can you help me improve my NPS score? How can you help me reduce redundancies in efforts and overhead? How can you help me look more like that company over there?" Once you've gleaned that information, then you can say, "Here's how content strategy or content design specifically can help you achieve that."

Another important thing to understand is that sometimes leadership, at least at the executive level, just doesn't care. In that case it's more effective to communicate with stakeholders instead of going straight to the top to get alignment on the work that needs to be done. "The fact of the matter is more often than not, content strategists, and even directors of content strategy don't have the opportunity to engage the C-Suite. It's more about when do you engage your boss's boss? *That's* the conversation. If you're advocating for content strategy, that's where you start."

There *is* one scenario where the C-suite is likely to engage, and that's when you're talking about where a content strategy practice should sit. Things can get political in larger organizations where you have content functions sitting across many different areas in terms of who owns what, including things like content management services, content marketing services, and so on.

It's particularly true when it comes to getting additional headcount, and when you've laid the groundwork to demonstrate the value of the practice. "That's the point where you make your case to your boss's boss, and then have your boss's boss come sit in with you when it comes time to make the case to leadership. Don't ever just walk in there alone. You have to have sponsors who are above you."

The Punch List

Wherever you are in the practice-building process, learning how to articulate metrics and other practice intel in a way that resonates with leadership is important for ensuring practice longevity. Taking the following steps will help facilitate conversations with whatever level of leadership is appropriate in your situation or organization:

- Understand the structure and supporting processes and procedures in your organization. Is it traditional and top-down, or is it autonomous? Or perhaps you're a team of self-organized content pros seeking to level up and form a practice.

- Figure out a phased approach to engaging leadership and when it's best to engage them in order to get maximum traction from those points of engagement.

- Learn the lingo of leadership or decode whatever flavor of acronym soup that might be used at your agency, with your clients, or within your organization.

- Consider using the SWOT analysis as a way to communicate effectively with key leaders and show how the practice can address each of the four parts of analysis to help reach organizational goals.

Use the information contained in this chapter at the time that best fits your circumstances and your organization—either before you start building, or as you reach important milestones along the way—to help you articulate to key leaders and stakeholders how the practice can help meet organizational goals.

CHAPTER 10

The Final Walkthrough

A fter a commercial structure or a new home has been built—when improvements or modifications are made to a space within an existing structure but before the space can be occupied—building inspectors conduct a final inspection of the space to ensure that it's been built to code and is safe and ready to move in. At that time, the tenant or occupant has a chance to conduct a final walkthrough to identify any last-minute issues that need attention or repair before moving in.

To ensure consistency, and to be certain inspections are thorough and complete, building inspectors, realtors, and other interested parties use a checklist similar to the one shown in Figure 10.1.

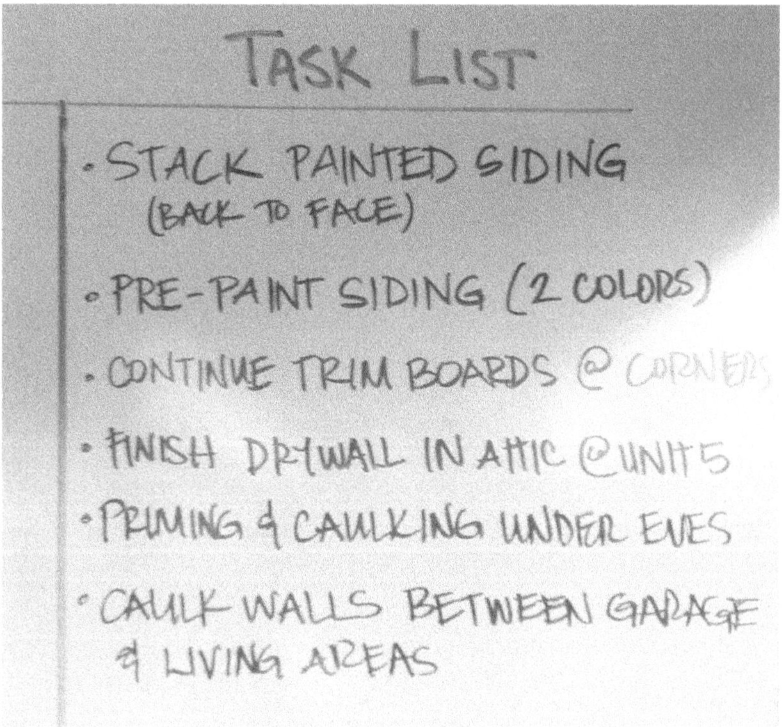

TASK LIST

- STACK PAINTED SIDING (BACK TO FACE)
- PRE-PAINT SIDING (2 COLORS)
- CONTINUE TRIM BOARDS @ CORNERS
- FINISH DRYWALL IN ATTIC @ UNIT 5
- PRIMING & CAULKING UNDER EVES
- CAULK WALLS BETWEEN GARAGE & LIVING AREAS

FIGURE 10.1

This checklist from a Habitat House home build helped a team of volunteer builders stay on task and ensured that no steps in the building process were missed.

This chapter provides you with a high-level checklist to help facilitate your final walkthrough of the practice you've built. This list will help ensure that you've covered each component of the Content Strategy Practice Blueprint introduced in Chapter 1, "The Content Strategy Practice Blueprint," and that you've implemented any key supporting processes or procedures that are critical to practice sustainability.

1. **Make the business case.** (Chapters 1-5)

   ☐ Identify the need for building a practice.

   ☐ "Soil test" the environment where you've built your practice to ensure that you can lay a firm foundation.

   ☐ Review the blueprint components to get clear on the type of practice you're building.

   ☐ Communicate the value of the practice to others.

   ☐ Create or curate case studies to demonstrate that value with real-world examples.

2. **Build strong relationships with cross-functional teams.** (Chapter 2)

   ☐ Create a list of cross-functional disciplines and teammates that are crucial to alignment and to create a strong foundation.

   ☐ Identify departmental partners beyond UX (or if applicable, DesignOps), who are crucial to alliance-building and maintaining a strong core.

3. **Create frameworks and curating tools to build with.** (Chapter 3)

   ☐ Involve cross-functional teammates in the creation of an end-to-end process framework.

   ☐ Identify roles and responsibilities, as well as critical handoffs within the process framework.

   ☐ Evaluate tools for use within the practice at the project or client level.

4. **Rightsize the practice to meet client or project demand.** (Chapter 4)

   ☐ Assess practice tools and structure durability to ensure that the practice can handle additional workloads.

   ☐ Revisit relationships with cross-functional teammates and departmental partners to share intent to scale the practice and identify process gaps and shore up any weaknesses in practice structure.

☐ If agency-based, proactively identify potential client projects to position the practice to scale in order to take on more work.

☐ If in-house, review product backlogs and roadmaps to identify upcoming projects where the practice can add value.

5. **Establish meaningful success measures.** (Chapter 5)

☐ Collaborate with colleagues and multidisciplinary partners to establish meaningful ways to measure practice success.

☐ Document success measures on a project tracker or in a similar repository.

☐ Share success measures with teammates, stakeholders, and at the appropriate level of leadership in your agency or organization.

Remember, you can always refer to the chapter on retooling (Chapter 7, "Retooling") at any point in the practice-building process to ensure that you have the right tools to maintain and scale the practice. And the chapter on scaffolding for sustainable growth (Chapter 8, "Scaffolding for Sustainable Growth") provides methodologies and approaches for testing and improving content strategy literacy and maturity across your agency or organization.

There is one final step in the practice-building process that's vitally important that you shouldn't overlook: the ribbon-cutting celebration! Not only does this last step signal to others that the practice is open and ready for business, but it also gives you a chance to step back, take a deep breath, and be proud of all that you've accomplished.

INDEX

build phase

of content strategy, 128

documents/deliverables, 130

of enterprise content strategy, 65–66

building a practice, 2–5, 18–21

building and construction. *See* construction metaphor

building materials. *See* process framework

building up and building out. *See* expansion: building up or building out

Burch, Noel, 159

burnout, 121

business case, making, 8–9, 10–11

business goals, quantifiable, 16–17

business objectives, in content strategy kickoff, 127

business speak, language of, 186–190

C

cadence establishment, 39–40

case studies, as method to maintain practice maturity, 163

Casey, Meghan, 6, 54, 130, 131

The Content Strategy Toolkit (Casey), 6, 54, 130

Managing Enterprise Content: A Unified Content Strategy (Rockley), 6, 89

Centers for Disease Control and Prevention (CDC), 42

change orders, 190

charter, for Community of Practice, 42, 45

check-ins

retooling, 145

status, 121–123

ClearVoice, 166, 169

Collins, John, 89–91

Communities of Practice (CoPs)

for content strategy, 44–46

online resources, 43

practice charter, 41, 42, 45

in structural alignment, 40, 42–43

comparative analysis, in rightsizing the practice, 14

competence, levels in understanding, 159–161

compression

as construction term, 4

testing the process framework, 66, 69–70

conferences for content strategy, 113

conscious competence, 159–160

conscious incompetence, 159–160

construction metaphor, 2–3

breaking ground, 3–4, 8

expansion, 80

final walkthrough, 198

inspection phases, 180

loads on the core, 118–119

measurement, 100

scaffolding, 156

structural durability, 12, 81

tension and compression, 4, 66–67, 69–70

home page with contest promotion, 72–74, 185

horizontal loads, 118

"How to Begin Designing for Diversity" (Gao & Mantin), 135

human-centered design (HCD)

    building cross-functional alliance with, 34

    Community of Practice for, 44

Humphrey, Albert, 167

I

implement phase, of enterprise content strategy, 64-65

inclusion, compared with diversity, 135

inclusive design, 134–135

information architects, building cross-functional alliance with, 35

instructional designers, building cross-functional alliance with, 149, 150–151

Intuit, 104–105

J

job site. *See* notes from the job site

journey mapping

    expanding practice operations, 83–84

    measuring success, 102

K

Kao, George, 182

key performance indicators (KPI), 188

kickoff for content strategy, 127–128

Knapp, Jake, 184

knowledge base of content strategy, 145

L

language of business speak, 186–190

leadership engagement, 179–195

    change orders, 190

    how and when, 180–181, 194

    importance of, 186

    language of business speak, 186–190

    organizational structures and presenting the practice to leadership, 181–185

    punch list, 195

    SWOT analysis, 192–193

Levy, Matthys, 118

Liminal Thinking (Gray), 133

loads on building cores, 118

localization efforts, 109

Løvlie, Lavrans, 85

lunch-and-learns, 107, 145

M

maintenance phase

    of content lifecycle, 130–131

    of content strategy, 129

    documents/deliverables, 130

Malouf, Dave, 131–132

Managing Enterprise Content: A Unified Content Strategy (Rockley), 89

notes from the job site

aligned for success, 67

case study, home page promotions, 72–74, 185, 191

contending for content strategy, 176

content strategy Community of Practice (CoP), 44–46

design sprints, 184–185

growing pains, 94–95

identifying the deciders, 176

passing inspection, 71

a practice in need of a plan, 20–21

starter set of tools, 61

SWOT analysis, 166

tipping point, 158

turf wars, 106

unrealistic expectations, 41

when the CEO met SEO, 191

writing and speaking, 114

O

objectives and key results (OKR), 189

office hours

in building a team, 19

in measuring success, 103, 107

online directory, 72–74, 185, 191

opportunities, in SWOT analysis framework, 167–169, 192–193

optimize phase, of enterprise content strategy, 64–66

organizational structures, and presenting the practice to leadership, 181–185

ownership responsibilities, 67

P

pain points, related to content, 33

partners, departmental, in journey mapping, 84. *See also* cross-functional teams

people, investing in for core maintenance, 120–126

Balance Score, 124–125

burnout, 121

self-care, 121–124

status check-ins, 121–123

vulnerability as power, 126

people/practitioners. *See* practitioners, content strategy

persistent principles, 22, 27, 43

phases of content strategy, 128–129

plan

in enterprise content strategy, 65–66

as phase of content lifecycle, 130–131

playbook. *See* practice playbook

Polaine, Andy, 85

Poole, Henry, 124–125

"The Power of Vulnerability" (Brown), 126

practice capabilities, as practice playbook component, 148

practice charter, 41, 42, 148

practice-level measures, 100

practice playbook, 102, 146–149

practice roadmap, 102, 143–146

practice types. *See* enterprise practice
   type; mid-sized practice type; solo
   practice type

practitioners, content strategy

   adding or leaving, as time to
      retool, 142

   content roles and responsibilities,
      89–91

   front-end and back-end, 88–89

   investment in people, 120–126.
      See also people, investing in for core
      maintenance

   in paths to expansion, 87–93

   as reinforcements from SWOT
      analysis, 169

principles, persistent, 22, 27, 43

principles for content, as practice
   playbook component, 148

process documentation, as practice
   playbook component, 148

process framework, 51–75

   content lifecycle and, 53–54, 58

   creation of, step by step, 58–60

   end-to-end mapping, 52–54

   importance of, 12, 140

   measuring success, 101

   phases of, 55–58

   as practice playbook component, 148

   punch list, 75

   streamlined or grow to fit, 70–71

   testing the framework: tension and
      compression, 66–70

   tools (books) for use, 60–66

processes, identifying, 86

product managers, building cross-
   functional alliance with, 35–36

profitability, analyzing, 86

project-by-project, content metrics, 100

project development phases, 55–58

project management, as phase of
   process framework, 57

project managers, building cross-
   functional alliance with, 36

project summaries, as practice playbook
   component, 148

project tracker, in measuring success,
   108–110

promotional modules on home page,
   72–74, 185

proposal, as phase of content
   strategy, 128

psychological safety, 121, 122–123,
   134–135

public speaking, 112–114

publish phase, of enterprise content
   strategy, 65–66

Q

quantifiable business goals, 16–17

R

Rach, Melissa, 6, 62

RACI, decision-making framework, 171,
   172–173

 Rosenfeld®

Dear Reader,

Thanks very much for purchasing this book. There's a story behind
it and every product we create at Rosenfeld Media.

Since the early 1990s, I've been a User Experience consultant, conference
presenter, workshop instructor, and author. (I'm probably best-known
for having cowritten *Information Architecture for the Web and Beyond*.) In
each of these roles, I've been frustrated by the missed opportunities to
apply UX principles and practices.

I started Rosenfeld Media in 2005 with the goal of publishing books
whose design and development showed that a publisher could practice
what it preached. Since then, we've expanded into producing industry-
leading conferences and workshops. In all cases, UX has helped us
create better, more successful products—just as you would expect. From
employing user research to drive the design of our books and conference programs, to working closely with our conference speakers on
their talks, to caring deeply about customer service, we practice what we
preach every day.

Please visit rosenfeldmedia.com to learn more about our **conferences**, **workshops**, **free communities**, and **other great resources** that
we've made for you. And send your ideas, suggestions, and concerns my
way: louis@rosenfeldmedia.com

I'd love to hear from you, and I hope you enjoy the book!

Lou Rosenfeld,
Publisher

RECENT TITLES FROM ROSENFELD MEDIA

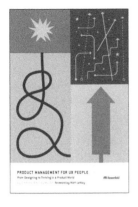

PRODUCT MANAGEMENT FOR UX PEOPLE
From Designing to Thriving in a Product World

LIFE AND DEATH DESIGN
What Life-Saving Technology Can Teach
Everyday Designers

SURVEYS THAT WORK
A Practical Guide for Designing and Running Better Surveys
by CAROLINE JARRETT

Get a great discount on a Rosenfeld Media book:
visit rfld.me/deal to learn more.

SELECTED TITLES FROM ROSENFELD MEDIA

CONVERSATIONS WITH THINGS
UX Design for Chat and Voice

WRITING IS DESIGNING
Words and the User Experience

LIVING in INFORMATION
Responsible Design for Digital Places
JORGE ARANGO

DIGITAL REALITY CHECKS 1
Digital and Marketing Asset Management
THE REAL STORY ABOUT DAM TECHNOLOGY AND PRACTICE
by THERESA REGLI

MANAGING CHAOS
Digital Governance by Design
by LISA WELCHMAN

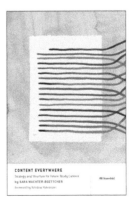

CONTENT EVERYWHERE
Strategy and Structure for Future-Ready Content
by SARA WACHTER-BOETTCHER

View our full catalog at rosenfeldmedia.com/books

ACKNOWLEDGMENTS

To my family:

Dakota, my son, and my biggest fan. From making dinner and taking care of the house, to sharing words of encouragement, you made it possible for me to focus on getting this work out and into the hands of those who will benefit from it.

My momma, Joyce. Your prayers, love, and support have guided me throughout this journey.

My "sissy," Barbara. Your encouraging texts, calls, photos, and GIFs kept me going on days when all I wanted to do was go back to bed.

To my friends:

Dawn Clark Archer, for always making time for coffee, brunch, and silliness, and for always reminding me that I am capable of doing epic things.

Louise McCarthy, for being my accountability partner on this journey, and for stoking the fires that kept me marching toward completion (before my deadline!).

Jonie Thomas, for always being a mirror of excellence, an icon of style and substance, and for never letting me forget my worth.

Rhea Weng, for suggesting that I write a book, well before I was ready to embrace it.

My angels, Denise Arndt and Soraya Medina, two fierce women who left this realm much too soon. At different times in my life, you both saw my potential for doing the hard things well, long before I was ready to accept and acknowledge it.

To my professional posse:

Kristina Halvorson: Your book was among the very first in my practice-building toolkit, and it was a career game changer for me. I am honored that you gave the most precious commodity of all—your time—to write the foreword for my book. Thank you!

Danielle Barnes, CEO of Women Talk Design, for holding space for underrepresented women and non-binary folks to become public speakers; for finding me on LinkedIn and inviting me to speak, and to become a WTD member; and for introducing me to Lou Rosenfeld.

To the people who so generously took the time to read and review this manuscript: Amy Mihlhauser, Lisa Welchman, and Andy Welfle. Your feedback and insights helped me write a much better book.

The voices of many others have helped to make this book diverse and inclusive, through their shared stories and lived experiences: Barnali Banerji, Kristina Halvorson, Jen Schmich, Andy Welfle, Candi Williams, and Berni Xiong. Thank you!

A huge shout out to those who made the time to read the manuscript with just weeks left before going to press: Meghan Casey, Aladrian Goods, Justin McKinley, David Dylan Thomas, and Sara Wachter-Boettcher.

Louis Rosenfeld, publisher, and book proposal groomer extraordinaire. It just dawned on me that I have no idea how it is that you asked Danielle to introduce us. And that's OK. I'm just *really* glad that you did. I will always appreciate how you helped me make my proposal come to life and evolve into this book.

Marta Justak, my editor, and friend. You put up with my initial hesitancy, my rambling introductory clauses, and many (many!) changes in the order of things as they appeared in this book. I promised myself that I would keep a beginner's mindset throughout this process, and I would listen to every morsel of feedback you provided so I could write the best book possible. We did it!

To everyone else at Rosenfeld Media who worked behind the scenes to accommodate an accelerated publishing schedule, answered my many questions about everything from the website to providing promo codes to share at conferences, and all of the work in between: *Thank you!*

And lastly, to my extended LA area DJ and dance family, for the music and the movement, as well as the shenanigans and the soundtracks, especially during COVID. Whether you streamed a show or masked up for an in-person event, your artistry helped me to find and share my own creative expression through this book.

ABOUT THE AUTHOR

Natalie Marie Dunbar is a UX-focused content strategist with a unique blend of skills as a journalist, content writer, and user experience researcher. Taken together with her curiosity for technology and her passion for engaging consumers, Natalie excels in balancing the creation of delightful user experiences with strategic content that supports the needs of a business or organization.

Natalie has worked in various roles as a content writer and strategist for brands that include Anthem, Farmers Insurance, Kaiser Permanente, Walmart, and YP.com. She's also produced original content for federal agencies that include the Animal and Plant Health Inspection Service (APHIS), Centers for Tobacco Prevention (CTP), the Food and Drug Administration (FDA), and the Veterans Administration (VA).

When she's not herding content, Natalie teaches private yoga, sharing the benefits of Hatha Yoga as a peaceful yet powerful way to reconnect mind, body, and spirit. In addition to her 200-hour training, Natalie holds certifications in Accessible Yoga, Curvy Yoga, Trauma-Informed Yoga and Yoga for All, and teaches body positive, all-inclusive yoga, for all shapes, sizes, and abilities.

Natalie lives in Pasadena, CA, with her son, Dakota, and fur kid, Gia.

Her "grand puppy" Occhilupo (a.k.a., Occhi), crossed the Rainbow Bridge as this book was coming together.

Along with Gia, Occhi was a constant companion to Natalie during the writing process.

Lightning Source UK Ltd.
Milton Keynes UK
UKHW020723290522
403667UK00006B/119